# Advance Praise for Ann Kirschner's *Lady at the O.K. Corral*

"Ann Kirschner brings a fresh, lively perspective to one of the great stories of the American frontier. *Lady at the O.K. Corral* reveals a fascinating intersection of Jewish history and the Wild West; its engaging narrative both celebrates and demystifies a legendary time and place."

—Julie Salamon, author of *Wendy and the Lost Boys*

"In this remarkable feat of historical sleuthing, Ann Kirschner coaxes the stubbornly evasive Josephine Marcus Earp out from behind the shadow of her famous partner, painting a vibrant portrait of an uncommon couple whose love for one another and shared thirst for adventure took them to the farthest reaches of the Wild West during its blustery boom times. Thanks to Kirschner's exhaustive research and fluid pen, Josephine Earp joins the ranks of Jessie Benton Frémont, Elizabeth Custer, and other spirited nineteenth-century women who defied custom to forge their own lives and shape the legends of the men they loved."

—Bruce J. Dinges, Arizona Historical Society

"A tour de force in the detective work of biography, Ann Kirschner's *Lady at the O.K. Corral* writes Josephine Sarah Marcus Earp back into American history. Mining unpublished manuscripts, personal letters, diaries, and court documents, Kirschner's fine narration tells the real story of the woman behind the man."

—David S. Ferriero, former director of the New York Public Libraries

"Ann Kirschner delivers a frontier story for the ages—part Unsinkable Molly Brown, part Mama Rose, part Queen Esther, the story of Josephine Earp proves that even the best lawman in the Wild West needed a good woman to stand beside him, as improbable as their romance was, and as riveting a read as this book most certainly is."

—Thane Rosenbaum, author of *The Golems of Gotham* and *Payback*

"Thanks to Ann Kirschner's brilliant *Lady at the O.K. Corral*, we finally have the definitive story of Josie Earp, a key player not only in the events leading up to and after the infamous shootout, but in crafting much of the mythology that's been widely accepted ever since. This is a must-read book for anyone who loves narrative nonfiction, or simply enjoys a hellaciously well-told tale."

—Jeff Guinn, author of *The Last Gunfight*

"Old West aficionados will find in this book a fresh account of the most famous of gunfights, but Ann Kirschner's engrossing biography of Josephine Marcus Earp offers much more. The life of Josephine that unfolds so vividly in these pages is as colorful and complicated as that of Wyatt, and as the reader will discover, hers was the more remarkable journey."

—Stephen Aron, UCLA and Autry National Center

"With a passion for research and an engaging flair for prose, Ann Kirschner has composed a biography of Josephine Marcus Earp that is a pleasure to read. No previous account has equaled in depth and understanding Kirschner's portrayals of Josephine and Wyatt—their families, their diverse associates, their lives in a rapidly changing American West."

—Harriet Rochlin, author of *Pioneer Jews: A New Life in the Far West*

"*Lady at the O.K. Corral* is the remarkable true story of Josephine Earp, who lived through the transformation of the western frontier from gold rush boomtowns to the back lots of Hollywood. In this vivid tale of romance and high drama, Ann Kirschner reveals the dark secrets of Wyatt Earp's past and Josephine's Jewish immigrant family."

—Abigail Pogrebin, author of *Stars of David*

"In a great piece of historical detection, Kirschner brings to life a woman who had previously been just a footnote, just an oddity. This book brims with the vibrancy of the Arizona Territory and situates the daughter of Polish Jewish immigrants into the rough and tumble of a half century of American life. This is a story that has never been told and that is just fine. It awaited Ann Kirschner's imagination, research, and sweeping prose."

—Hasia R. Diner, author of *A Time for Gathering*

# LADY AT THE O.K. CORRAL

ALSO BY ANN KIRSCHNER

*Sala's Gift*

# Lady
## AT THE
# O.K. CORRAL

The True Story of
JOSEPHINE MARCUS EARP

## ANN KIRSCHNER

HARPER

www.harpercollins.com

HarperCollins books may be purchased for educational, business, or sales promotional use. For information, please e-mail the Special Markets Department at SPsales@harpercollins.com.

FIRST EDITION

Library of Congress Cataloging-in-Publication Data has been applied for.

ISBN: 978-0-06-186450-6

13 14 15 16 17   OV/RRD   10 9 8 7 6 5 4 3 2 1

FOR JOEY, MY FIRST HERO

Whatever you choose, however many roads you travel, I hope that you choose not to be a lady. I hope you will find some way to break the rules and make a little trouble out there.

—*Nora Ephron, Wellesley College commencement address*

I don't have to speak, she defends me.

—*Robbie Robertson, "Up on Cripple Creek"*

# | Contents

Prologue:
In Which I Land on Planet Earp      1

**1**
A Jewish Girl in Tombstone      13

**2**
The Fourth Mrs. Earp      69

**3**
The Greatest Mining Camp the World Has Ever Known      89

**4**
Waiting for Wyatt      135

**5**
Josephine's Last Trail      177

**6**
Planet Earp      221

     Acknowledgments      239
     Earpnotes      243
     Notes      247
     Sources      263
     Photo Credits      271
     Index      273
     About the Author      290

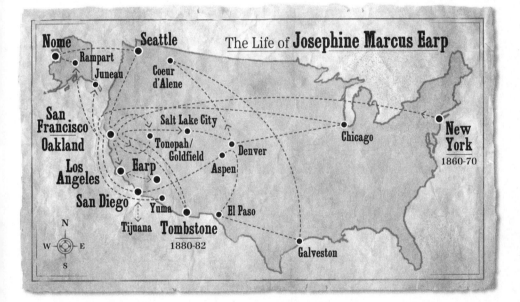

The Life of **Josephine Marcus Earp**

Nome
Rampart
Juneau
Seattle
Coeur d'Alene
San Francisco
Oakland
Salt Lake City
Tonopah/Goldfield
Denver
Chicago
New York
1860-70
Los Angeles
Earp
Aspen
San Diego
Yuma
Tijuana
Tombstone
1880-82
El Paso
Galveston

N
W E
S

## Prologue | IN WHICH I LAND ON PLANET EARP

*D*ID YOU know that Wyatt Earp was buried in a Jewish cemetery?"

Just hearing his name threw me back to my childhood in Jackson Heights, New York City, sprawled on the floor in front of a black-and-white television, watching Westerns with my big brother Joey, dressed up in his special shirt with braided trim and snazzy snap buttons and black cowboy hat and shiny gun in a faux leather holster slung around his hips. Joey and I tuned in and pretended to walk the streets of Tombstone every week, together with millions of Americans, young and old.

Joey was my hero, and Marshal Earp was his.

Brave, courageous, bold . . . and Jewish?

That's how it all started, just an innocent question from a friend who thought (correctly) that I would be intrigued by the incongruities between anything Jewish and anything Tombstone-ish.

This first burst of curiosity about Wyatt Earp's final resting place and religion was easily satisfied. I soon learned that Wyatt, the only man to emerge unscathed from the Gunfight at the O.K. Corral, was not Jewish but had lived with a Jewish woman for nearly fifty years. She buried him next to her parents and brother in a family plot at the synagogue-affiliated Hills of Eternity cemetery outside of San Francisco.

And that was my introduction to Mrs. Earp.

As children when Wyatt Earp ruled the airwaves, we didn't know that he had a wife, certainly not a Jewish wife from New York. I was a Jew from New York! So each revelation about a woman named Josephine Sarah Marcus evoked new images that made me smile, especially the thought of Wyatt Earp going home for chicken soup after a tough day fighting for truth and justice in the dusty streets of Tombstone, Arizona.

I quickly became far more interested in Mrs. Earp than in her famous husband. Contradictions piled up like a freeway collision. How had a beautiful girl from San Francisco via New York and Prussia ended up in Tombstone? While the rest of her immigrant family climbed out of poverty and into bourgeois respectability, why had Josephine run away? What inspired five decades of adventure-seeking that took her from the Arizona Territory to California, Nevada, and Alaska, and then, finally, to Hollywood?

The historical record on Josephine was thin. None of the early articles and books about Tombstone—and there were many—even mentioned her existence. On the other hand, the history of Wyatt Earp and the Arizona Territory sank deep roots into the American psyche. Books about Wyatt's exploits became best sellers, movies played to full houses, and historical reenactments drew large crowds. Wyatt Earp was big business, then and now. Try a random Internet search today, and the O.K. Corral will pop up in a motley assortment of links to political contests, basketball championships, and stock traders fixing interest rates. On Wikipedia, the story of Wyatt Earp is told in twenty-six languages, but Josephine was, until recently, nowhere to be found.

On October 26, 1881, Wyatt, his handsome brothers, and Doc Holliday strode down Tombstone's main street and confronted a quartet of murderous cowboys and thieving cattle rustlers near the O.K. Corral. Less than a minute and dozens of bullets later, three men were dead and three were injured. Wyatt Earp walked away without a scratch.

The gunfight never loosened its iron grip on the public's imagination. For more than a century it has remained an important symbol of American culture, evoked in the wake of every contemporary tragedy, from Columbine to Gabrielle Giffords, an enduring reference point for our national obsession with guns and violence and revenge. Along the way, "cow-boy" lost the hyphen and mutated from lawless marauder to the Maverick Man. Tombstone and the O.K. Corral became intertwined with other powerful themes of American history: industrialization, urbanization, seismic shifts of attitude toward gambling and alcohol and prostitution, and the end of the American frontier.

I thought I knew the story. But I knew only one version. The O.K. Corral is a story with a thousand variations. There is, for instance, the Marxist script that casts the Earps as the repressive embodiment of a capitalist regime, determined to impose law and order on the frontier to keep profits flowing. In the Manichean narrative, Wyatt Earp is a defining figure in the dynamic interplay of good and evil.

As Josephine and I became better acquainted, I wanted to hear the human version that restored to its center the real life of one woman caught up in the frenzied atmosphere of the most famous frontier boomtown. The Gunfight at the O.K. Corral was a love story, fought over Josephine Marcus, a woman of beauty and spunk barely out of her teens, escaping the restrictions of birth and seeking adventure, independence, and romance.

SHE HAD BEEN born on the wrong side of the tracks. For the rest of her life, Josephine would characterize her father as a "wealthy German merchant," but that was one of her cover-up tales. Her mother and father were from Prussia and subject to the deep-seated prejudices of German Jews against their Eastern European cousins. They had neither the connections nor the education nor the wealth to launch their daughter into the highly stratified German-Jewish aristocracy of nineteenth-century San Francisco.

Although she never experienced anything as dramatic as a spiritual crisis or loss of faith, Josephine was indifferent to being Jewish and believed that she could leave her family history behind. As Nora Ephron would memorably say of herself, Josephine was not in denial; she acknowledged that she was a Jew, but being Jewish was not in the top five things she wanted you to know about her. Her few nostalgic references to home were linked to her mother and some affectionate Yiddish blessings and songs. An accomplished cook, she preferred biscuits and other southern specialties to the schmaltz of her childhood. She relinquished all personal ties to Judaism or any organized religion, preferring a kind of loosely held pantheism that peeked out occasionally when she was overwhelmed by the grandeur of the Californian desert or the wild Yukon River. Wyatt Earp had more Jewish friends than she did. In the rough and paranoid years of her later life, she stooped to label her enemies "kikes" or "sheenies."

And yet when she confronted her greatest crisis, she turned back to her family's faith and buried her husband among the Jews of San Francisco.

Josephine was street-smart and intuitive, playful and fearless and headstrong. A mixture of naive and sophisticated, she had no interest in politics, and no one remembered her picking up a book or a newspaper. She yearned to be a lady, while spending much of her life among gamblers and gunfighters. She was a man's woman, with a handful of female friends. None of her husband's status as popular culture hero would rub off on her: there would be no television shows or theme songs about her, no lunch boxes with her image.

Josephine was beautiful. With her spectacular figure, piles of dark hair, and strong features, she must have looked like a Victorian-era Penelope Cruz, commanding every male eye in Tombstone. As an older woman, her silhouette still amazed Grace Welsh Spolidoro, a close family friend but no fan of Josephine's: "Her bosoms came in the front door before her body did! She was humongous, and she was a small lady. Little hips, little legs, but those things were the biggest things I ever saw."

Josephine was bold, and she was funny. She had a laugh that was described as "the tinkling of champagne glasses" and a gift for mimicry that delighted children. She had a nose for adventure, and feared little other than ennui—and snakes. She remained close to her family, always a caring and attentive aunt and great-aunt, but she sought a permanent escape from the narrow, predictable future that she foresaw for herself, caught up in the German-Polish divide.

Unlike her sisters, she chose a life of excitement over a home and children. The girl who ran away from home at eighteen lived to be an old woman of eighty-four without ever having a single permanent address. Or a legally binding marriage contract. Josephine never said, "I do," only "I will." But her lifetime partnership with Wyatt Earp was indeed a marriage, with the power to withstand the tempests of infidelity, the pressures of sudden wealth and slow poverty, and the anxieties of aging and illness.

Once hooked on uncovering her story, I acclimated myself to a strange new place I called Planet Earp. Here, Wyatt Earp had his passionate supporters and detractors, and everything about Josephine was controversial, down to her age and name. Although her parents and her siblings had undisputed birth dates, Josephine remained unsure as to her own, not only because it suited her to be as young as possible, but because no public record existed to confirm the date that she believed to be correct: June 2, 1860. A sizable group of authenticated photographs testified to Wyatt Earp's undeniable good looks at any age, while there was not a single undisputed photograph of young Josephine, only ones in which she looked more Sophie Tucker than Penelope Cruz.

Even her name was contentious. From 1881 until 1929, Wyatt mostly called her Sadie, a nickname from her middle name, Sarah. After Wyatt's death, she could not bear to be Sadie and insisted on Josie only. But some denizens of Planet Earp mocked her preference of Josie as pretentious or a vain attempt to hide her past. I discovered that I could stumble over some fatal tripwire if I used the wrong name. Heads would shake with disapproval. Sources would dry up. Conversations would end.

And so I called her Josephine.


JOSEPHINE MARCUS EARP wasn't the only woman erased from western history; with few exceptions, Planet Earp was inhabited only by men. The early chroniclers of frontier lore, writers like Walter Noble Burns, Frederick Bechdolt, and William Breakenridge, seemed to have no wives, mothers, daughters, lovers. At least, they didn't write about them. When Western writers did insert an occasional dance-hall girl back into the landscape, her portrayal was often crude and lifeless. And more recently, when modern historians did begin to pay attention to Wyatt's women, the stories were often inaccurate and filled with ugly caricatures. Whenever I heard Josephine described as "shrewish," I suspected that the writer really meant "Jewish."

Josephine had the right to her own story. Feminist scholar and writer Carolyn Heilbrun might have been writing about Josephine when she pointed out that denouncing women as shrill or strident—accusations often made about Mrs. Earp—was another way of denying them power. Josephine had the additional complication of belonging to an immigrant Jewish family. She aged into an America that would experience a wave of anti-Semitism, especially in the 1930s, when Josephine was at her least likable, a querulous old lady recovering from the death of her lifelong partner.

Would Josephine be furious to have been all but rubbed out of the pages of history, or relieved?

I wanted to answer that question, not to provoke a discussion of misogyny or anti-Semitism among Western writers and historians, though I could argue that one was long overdue, but because I had news: there was a woman at the O.K. Corral.

Only a handful of people even knew that she was there, and few recognized that she was at the apex of a love triangle as the former fiancée of the Cochise County sheriff, the champion of the cowboys' cause and Wyatt's political rival. She fell in love with Sheriff Johnny Behan, only to discover that she had chosen poorly. He wooed her with promises of marriage, but she would soon find herself alone in the hostile climate of a frontier boom-town, where a single woman in need of food and shelter could easily find herself working as a prostitute.

None of her contemporaries knew why Josephine Marcus came to Tombstone, or why she left. As eyewitnesses died off, it became harder and harder to follow the trail of broken promises and festering secrets. But fascination with Wyatt Earp and the West increased. He became famous as the iconic American lawman, a stoic, ambiguous figure at the heart of the ultimate American morality play, sometimes wearing a badge and defending the law, and sometimes in pursuit of vigilante justice. "Manhood extreme," as his friend Bat Masterson described him, this improbably handsome man attracted a remarkable variety of loyal friends, from Doc Holliday to Senator George Hearst to Endicott Peabody to Tom Mix and William S. Hart. Successive generations used him as an archetype from which to fashion contemporary versions of a lawman who stalked the streets of gangland Chicago, the jungles of Vietnam, and more recently, zombie-postapocalyptic landscapes and alien battlefields of outer space.

Before the archetype, there was a man. The real Wyatt Earp lived with the same woman for most of his adult life. Little has been known about that woman or her critical role in shaping the life and legend of Wyatt Earp. Until now, hers was just one more untold tale of the women of the West. Yet our national narrative is flawed, our canvas incomplete, until Josephine Earp makes her entrance, which is the purpose of this book.

Josephine spent decades shaping the public face of Wyatt Earp. Her motivation was complex, even contradictory: to give immortality to the legend of Wyatt Earp and to erase the dark shadow that Tombstone cast over their life. She dreaded anything that came close to the unsavory, violent truths that were the foundation of their long life together. If her pleas and tears did not suffice, she threatened lawsuits against anyone who threatened to reveal the secrets of their past. She and Wyatt came to dread the very mention of Tombstone—yet Tombstone was undeniably the wellspring for her subsequent adventures, and Wyatt's everlasting fame. "Surely there is something more pleasant to talk about," Wyatt would say to inquiring folks. Josephine hid her tracks so well that when her own niece visited Tombstone, she called it "Uncle Wyatt's old hang out," with nary a word about her aunt.

When Josephine finally agreed to tell her story, she recruited two Earp relatives, Mabel Earp Cason and Vinnolia Earp Ackerman, but even to them, she could not bear to reveal the truth.

Never sure whether she should speak or be silent, Josephine destroyed one version of her memoir and put a curse on anyone who dared to publish it.

The result was that for more than a hundred years, most of what was written about Josephine Sarah Marcus Earp was a lie—and no one told more whoppers about Josephine than she did herself.

What was she hiding?

CURIOSITY BECAME OBSESSION. I fell in love with Josephine, the flamboyant, curvaceous Jewish girl, the restless romantic with a persistent New Yawk accent. She began her nomadic life on horseback and in stagecoaches, and would later travel by railroad and eventually ride in her own car, always moving, always looking for the next destination, making and losing fortunes, driven by what Wallace Stegner called "the incurable Western disease." She would be at home in the deserts of the American Southwest and the boomtowns of the Alaskan gold rush, in the opulent hotels of San Francisco in the Gay Nineties, in rough mining camps, gaudy gambling casinos, racetracks, and boxing arenas, and finally she would be received among the royalty of Hollywood. As Mabel Cason's son told me, she was the most aggravating, frustrating, interesting woman he had ever met, as far from "plain vanilla" as you could get.

I set out to find answers to the mysteries of the woman at the O.K. Corral. I wanted to understand how a woman could survive in the crazy boomtowns that were Tombstone, San Diego, and Nome. How did she fuel her bottomless capacity for self-invention? What sustained her life-long partnership with a man of uncommon charisma and complex hero-ism? Her life enveloped me as the untold tale of a private woman every bit as interesting as her husband. I wanted to understand this woman of con-tradictions: the young runaway who stayed close to her family, the Jew who shunned other Jews, the rebel who could never make up her mind about whether she should be the lady or the tramp. Josephine sought to cover herself with a cloak of respectability, but she couldn't quite make it stretch over a life that was as unpredictable and dynamic as the American frontier itself.

Wyatt and Josephine lived together for forty-seven years. She drew her strength from him, but she was the one who managed his business, signed his letters, and entertained his friends. He was buried with tears and eulo-gies and coast-to-coast headlines, while she died practically destitute and friendless.

All of which brought me back to the gunfight of October 26, 1881.

I WENT TO TOMBSTONE. I walked its streets, thinking about its pecu-liar place in history, how it was at one time destined to become the capi-tal of Arizona, only to be practically abandoned after its precious metals were depleted, the bankers, brothels, and booze long gone. The mines that once operated twenty-four hours a day were open only as tourist attractions. At the annual Helldorado festival, created in 1929 to keep Tombstone alive as a business and tourist center, the streets filled up with thousands of Wyatt Earp wannabes, riding in stagecoaches, waiting for hourly reenactments of the gunfight, and ready to twirl a bushy mustache for photographs. Lost in the town's tacky present was the drama of the once sharply drawn factions of another century: Confederate loyalists versus abolitionists, rural cowboys and ranchers versus townsmen and capitalists. They voted for different parties, and they read different newspapers. Even their clothes were distinct, with the urbane Earps on one side in long black coats, and the cowboys on the other side in their farm clothes and bandannas. They collided at the O.K. Corral, a place named in honor of President Martin Van Buren, nicknamed "Old Kinderhook" after his hometown in upstate New York.

So what about the other shootout that day, the one where the central character was Josephine and the prize was love? How did this nice Jewish girl end up at the center of an iconic event with one of America's greatest folk heroes? What drew her into a love affair that was at odds with every value and experience of her immigrant family?

When I went searching for Josephine, I found myself caught up in an adventure in which the detective work of a biography brought me in contact with an extraordinary cast of characters. There are scores of professional and amateur historians who have made lifelong studies of the life and legend of Wyatt Earp and his family. Most of them are honest brokers of western history. I met heroes, who cheered me on and helped me gain access to important new sources and archives. I encountered some villains, who were not so fond of women or Jews, so they wished nothing but obscurity—or trouble—for Josephine and me. And sometimes it was hard to tell the heroes from the villains. You'll meet some of both in the last chapter.

I negotiated some tricky pathways that admitted me to important new sources, including the unpublished manuscript of Josephine's memoir, *I Married Wyatt Earp*, written with her collaborators Mabel Earp Cason and Vinnolia Earp Ackerman. My understanding of Josephine's life was profoundly enhanced by the privilege of reading this memoir, now in the archives of the Ford County Historical Society in Dodge City, Kansas. However, Cason and Ackerman had large gaps and inconsistencies in their story, usually because of Josephine's deliberate misdirection, which also made it impossible for them to finish their book. Their problems were compounded by the publication of *I Married Wyatt Earp: The Recollections of Josephine Sarah Marcus Earp* (1976), collected and edited by Glenn Boyer, a controversial book that kicked up a city-size cloud of intrigue over Josephine's life, as thick as the gun smoke that hung over Tombstone on that fall afternoon in 1881.

I wanted to clear the dust away and see Josephine with fresh eyes. For that, I had to peel back the layers of misinformation that hid the truth, going back to authenticated primary sources whenever possible. Historians must forever be grateful to Mabel Cason and Vinnolia Ackerman for saving their manuscript; to John Flood, for disobeying Josephine's orders to destroy her correspondence; and to Stuart Lake, whose archive at the Huntington Library is, as Jeff Morey has noted, his true legacy, every bit as important as his highly sensationalized biography of Wyatt Earp. I was the first to have access to the greatest number of Josephine's original letters, many of which are still in the hands of private collectors. Remarkable tape-recorded interviews with eyewitnesses were shared with me, many done in the 1960s, when recollections of Josephine were reasonably fresh. Walter Cason brought a bracing dose of realism and integrity to his memories of "Aunt Josie."

As I struggled to portray Josephine with all her paradoxes, I was inspired by the high standards set by the writer Mabel Cason. Weighing incomplete or contradictory evidence, I asked myself: What would Mabel do?

Josephine's biography would demand an understanding of the forces that inspired her life decisions—the social history that influenced her as she shaped Wyatt's role in the myth of the West. This would be her most lasting achievement. But I also sought to uncover the story of an American marriage, forged in the frontier. In these pages, as he was in real life, Wyatt Earp will never be far away from Josephine.

There was a woman at the O.K. Corral. That one minute of mayhem changed her life as inexorably as it did Wyatt Earp's.

It all started in Tombstone.

# 1 | A JEWISH GIRL IN TOMBSTONE

*L*ONG BEFORE you saw the city of Tombstone, you could smell it and hear it and feel it in your bones.

Josephine left San Francisco and set out for the Arizona Territory just before Christmas, 1880. She traveled most of the way by train, and then boarded a stagecoach in Benson for the last leg to Tombstone. From the windows of the stagecoach, Josephine admired the raw beauty in the shadows that shifted over the mountains and giant rocks. As they drew closer to Tombstone, her holiday mood shattered as she and other passengers bounced painfully across twenty-five miles of rough road that left them nearly suffocated, covered with dirty, nasty-tasting dust, and aching in every muscle and bone.

A year ago, she had been a teenage runaway, headed for the Arizona Territory with dreams of becoming an actress but already suffering pangs of guilt for leaving her mother. She was a pretty and vivacious young girl, small in stature but already showing signs of the voluptuous woman she would become.

Now she was entering Tombstone for the second time, engaged and about to join her fiancé. She had changed greatly between the two journeys. She had given up on acting, but not on romance. The most striking hallmark of her character, her love of adventure, had just begun to emerge. Josephine was well aware of her exotic appeal, especially the effect of her expressive brown eyes and womanly curves. She was a confident and charming young woman with full breasts and a narrow waist, a ripe beauty who drew men's eyes wherever she went. Her strong features and creamy complexion were set off by tumbled curls of dark hair, only now growing back after being chopped off the year before, during her brief stint as an actress.

Every mile she rode toward Tombstone took her farther from San Francisco and from her life as the daughter of Sophia Lewis and Hyman Marcus, Jewish immigrants who had arrived separately in New York around 1854, escaping the political upheaval and limited opportunities in their native towns in the province of Posen. A large Jewish community, Posen had been partitioned from Poland in the late eighteenth century and absorbed into Prussia, which would eventually become the core of modern Germany.

Prussian policy rewarded assimilation. Jews were encouraged to convert to Christianity with the carrot of citizenship and the stick of discrimination; many Jewish schools were closed down, and Jews were severely restricted in where they could live, work, and marry. Other Jews were emancipated in 1820—but not the Jews from the annexed region, who were denied citizenship for another generation.

Class distinctions evolved within the newly expanded population. Some stereotypes came from within the Jewish community itself. German Jews sought to project an image that was secular, sophisticated, and upwardly mobile, in stark contrast to poorer, less educated and worldly Prussian Jews who spoke mostly Yiddish and were more likely to stay stuck at the bottom of the social ladder. Craftsmen or small merchants, they were also most vulnerable to dislocation in a rapidly industrializing European economy.

In the minds of those given to labels, these Jews were less likely to succeed, more likely to be disparaged as "Polacks." It was a stubborn prejudice that would survive well enough to be transplanted into the soil of another continent.

Hyman Marcus had every reason to leave Prussia. His own father, the baker Moses Marcus, had probably received citizenship in 1834, but Hyman faced an uncertain future. He could follow his father's profession in their town or move to a larger city—perhaps Breslau or Berlin. Or he could start over in England or the New World. Like thousands of other Jews from Prussia, he chose America.

Hyman booked passage to New York, the most popular and the easiest destination to arrange, even for families or women traveling on their own. Soon after he arrived, he met and married the widow Sophia Lewis, who had emigrated a few years before with her daughter, Rebecca. Sophia and Hyman were married around 1855, and three children followed: Nathan in 1857, Josephine in 1860, and Henrietta in 1864.

Despite the strangeness of their new country, Hyman and Sophia were comforted by the presence of many other immigrants. They probably settled in the Five Points neighborhood of Manhattan's Lower East Side, where they would have found the streets filled with Jews speaking German and Yiddish, including many from Posen. The Jewish population of New York City more than tripled in the 1850s, reaching about 60,000 by the end of the decade. Hyman would have been relieved to see an abundance of jobs in food production, especially for bakers and confectioners. One local directory shows that there were nine Jewish bakers preparing "real kosher" matzos for Passover in 1859.

However, conditions in pre–Civil War years were primitive and dangerous. In Five Points, families crowded together in slums with backyard wooden privies. Outbreaks of typhoid were common. The city's highest mortality rates were of immigrant children under the age of five. There were public schools, but many immigrant children did not attend classes at all, and were taught at home, or worked side by side with their parents. New York City had a police department, but no professional firefighters.

The most influential forces in the community, next to the synagogues, were the ten Jewish newspapers that recorded births, deaths, marriages, political meetings, cultural events, and advertisements, vying with Civil War coverage, and with exciting letters from transplanted New Yorkers who had made their way to San Francisco. The West loomed large in the imagination of all Americans and must have been particularly irresistible for immigrant Jews who had come searching for the promised land of economic opportunity, social mobility, and political power, and had not found it in New York. Now San Francisco beckoned to them as a remarkable Jewish success story.

San Francisco sprang up as a child of the 1849 gold rush. When the output of California gold, so extraordinary in the early years, began to decline, San Francisco thrived because of its harbor and geography, evolving quickly from a village of shanties and wood-frame buildings to a metropolis with a diversified economy that could support a large workforce of skilled and unskilled laborers. San Francisco became the home of an affluent and influential Jewish community, which included retail businesses founded by Levi Strauss and Solomon Gump, as well as the founders of the Wells Fargo Bank and Lazard Frères. Although smaller in absolute numbers than the New York Jewish community, San Francisco boasted a similarly glittering cast of entrepreneurs in retailing and manufacturing, bankers and stockbrokers, and represented a wider range of Jewish economic interests, from fur trading in Alaska to wheat farming, sugar mills, and vineyards.

With arrivals from England, Australia, Holland, and Russia, as well as the large representation from Germany and Prussia, Jews created a vibrant cultural and religious life within the city, and constructed a sturdy social safety net for those less fortunate, including orphan asylums and services for the elderly. There were five Jewish newspapers, and Jews were active in political life, electing merchant Abraham Labatt as alderman for the new city of San Francisco in 1851. Considerable chest-thumping proclaimed San Franciscan Jews to be the most prominent and prosperous group of Jews in the world, so integrated into San Francisco's commercial life that steamer service—the city's vital link to the rest of the world—was suspended on the Jewish high holidays. An enthusiastic editorial in the *Daily Alta California* lauded Jews for being true Californians, and congratulated the non-Jewish Californians, since "no other part of the world can instance a similar act of liberality."

With this encouraging track record of tolerance, plus a temperate climate and booming economy, San Francisco had much to offer an immigrant Jewish family. Hyman Marcus had achieved only modest success as a baker, and he and Sophia were ready to leave the filth and the grinding poverty of Five Points. They had already made the far more difficult decision to leave Europe; the second leg of their family voyage at least would not require another new language and a new continent. It was not uncommon for young and energetic immigrant families to make multiple moves. In fact, rate wars between rival railroads and steamship companies made it actually cheaper for some families to move than to pay the rent.

The Marcus family made the bold choice to continue west across a continent that was still decades away from regular train service. Together with some 40,000 other people who would travel to California in the late 1860s, they arrived in San Francisco in time to be counted for the 1870 census. While Sophia Lewis's obituary would later identify her as a California pioneer who sailed around the Horn, the family most likely crossed the Isthmus of Panama, which replaced the dangerous eight-week-long all-sea voyage to become the major artery connecting the East Coast to California and the primary route for transporting gold, mail, news, and packages. Adult passengers watched the scantily dressed Panamanian natives through the train windows, marveled at the rain forests, and exclaimed at the deep canyons and unfamiliar flora and fauna.

Once the travelers reached the western coast of Central America, they boarded another overcrowded steamship. Josephine, then about eight years old, remembered almost nothing of the journey. Since she tried to eliminate any impression that her family was ever poor, she would have been silent anyway about an unpleasant three weeks in steerage.

The San Francisco that greeted the Marcus family was still recovering from the 1868 earthquake, but its economy was robust. The chaotic boomtown atmosphere of the gold rush years was gone; schools, community centers, and restaurants replaced brothels and saloons. The completion of the transcontinental railroad in 1869 would soon connect San Francisco even more to the outside world. The city was growing up, but it retained a strong entrepreneurial spirit.

Although Josephine's new neighborhood was a step up from Five Points, middle-class stability and status still eluded the Marcus family. Josephine would later claim that her father had a "prosperous mercantile business" that gave her a "comfortable and prosperous" home, but all evidence points to a more precarious existence. Hyman was still a baker. Local directories show that the family moved frequently—at least six times in their first ten years in San Francisco, all within the lower-class sections of Ward 4.

Not all Jews were so sure that the wealth and influence of San Francisco was a good thing. Among the most fascinating portraits of nineteenth-century San Francisco is that of Israel Joseph Benjamin. Born in 1818 in the province of Moldavia (now Moldava and part of Romania), Benjamin had some commercial failures before seizing on the idea that he might emulate the great medieval traveler Benjamin of Tudela and wander the world as an itinerant preacher and commentator on contemporary Jewish life. Supported by donations and hosted by curious local leaders, he traveled extensively throughout the Near East, North Africa, and Europe before sailing to the United States.

Benjamin saw much to admire in the thriving cultural and intellectual life of American cities. But he was unsparing in his contempt for their emphasis on commercial success and indifference to high culture and scholarship. Repelled by the materialism he saw, he concluded that neglect of learning was the source of all American misfortunes, not only those of Jews. "Will America be able to become a nation of princes without the education of a prince?" he asked rhetorically. In his view, young people who were brought up to be merchants, bankers, farmers, and mechanics were not destined to become independent, compassionate, or intellectual.

Benjamin devoted one of his longest stays to San Francisco. He observed that the pinnacle of society belonged to the Germans, and the lowest class rungs were the provinces of Polish Jews. He was amazed to discover that when a major fire destroyed two streets and eight large warehouses in San Francisco, the Eureka Benevolent Association, one of several Jewish social services organizations, turned its back on Polish Jews made homeless by the fire. Benjamin related the sad tale of a religious Polish Jew who applied for help to feed his many children. Refused by the board of directors—which included mostly Germans and Frenchmen—he went to an American Christian and told him the story; according to Benjamin, "that person went to the president of the society and shamed him into giving the Pole money."

SOON AFTER THE Marcus family arrived in San Francisco, Josephine's half sister Rebecca married Aaron Wiener, a clothing salesman who was also from Prussia. Josephine never forgot the sight of her father's stovepipe hat and long tails, her mother's wine-colored moiré, the bride's pale silk lavender dress trimmed with white satin, and her own yellow high-button shoes. She recalled her errand to purchase "diamond dust" to sprinkle over the bride's elaborate, perfumed coiffure. Young Josephine could not resist opening and sampling the precious powder, which spilled out of the package and blew away.

Josephine found the strict discipline of school disagreeable, and was not much of a student; her sister Henrietta, known to all as Hattie, was the family's star scholar. The sisters' education was, however, a sign that the family had finally risen to the middle class. Neither Josephine nor Hattie was put to work to support the family, nor is there evidence that Mrs. Marcus worked outside the home.

As a child, Josephine was infatuated with the stage; the abundance of San Francisco theaters and their low ticket prices made it a favorite pastime. Among her closest playmates were the Belasco daughters, whose brother David was already on his way to becoming one of the most famous playwrights of his era. It was a heady time for Jewish performers like Adah Isaac Menckens, who stunned even worldly San Franciscans with her onstage appearance in flesh-colored tights. Rebecca took Josephine and Henrietta to the opening of the Academy of Music, never dreaming that its owner, Elias Johnson (Lucky) Baldwin, would one day be Josephine's close acquaintance. San Franciscans also loved artists and art; one painting by the Jewish artist Toby Rosenthal was so popular that thousands of people lined up to pay twenty-five cents to see it. When it was temporarily stolen, people filed past to pay their respects to an empty picture frame (it was eventually returned).

Josephine's rebelliousness began to emerge in small ways, such as having her ears pierced by her Chilean classmate's *abuela*, though her mother had forbidden it many times in the past. Josephine dragged Hattie with her, taking advantage of the young girl's admiration and willingness to follow her older sister blindly. Looking back on her adolescent self, Josephine acknowledged that she needed discipline, but chafed against it. She contrasted the "tolerant and gay populace" to the "merciless and self-righteous" child-rearing philosophy embodied in the public schools.

Even if her father had risen more quickly beyond his humble origins and lowly profession, even if they'd had servants and membership in the right clubs, the Marcus family would still have been outsiders in the upper reaches of San Francisco Jewish society, where a less than perfect German accent signaled "second class." Everywhere around Josephine were intense signs of the social stratification that determined one's future. The odds were stacked against her: she was no longer the poorest of the poor, but she was not likely to win a German husband. She went to the wrong schools and was invited to the wrong parties. Nor did she have the talent or sheer will to break through all those barriers and still distinguish herself as an educator, lawyer, artist, or political activist, careers pursued by some Polish Jewish women in San Francisco.

Seeing the signs of a surging American economy and upward mobility all around her, a proud and energetic young woman like Josephine would have resented the assumption that she was inferior by birth. How ironic that in a city with an impressive lack of anti-Semitism, the Jewish community could be blamed for imposing antique prejudices on its own members. For some notable successes such as David Belasco or Gertrude Stein, San Francisco would be celebrated as the epitome of bohemian sophistication and freedom, exotic and freewheeling. For Josephine, San Francisco meant a predictable, dull life of lowered expectations.

To be one more struggling Polish Jew from an unremarkable family was simply not enough. It was adventure Josephine craved, and none of the Jews she knew were a match for her. So she would barely acknowledge her Jewish birth, and no Jewish organization would ever count her as a member. She would reinvent her parents as a wealthy German merchant and his proper German wife, who brought their daughter up to be a lady. Her status as a San Francisco Jew would figure only as a marker of the life she was about to leave behind.

Josephine Marcus would write another story for herself.


EIGHTEEN SEVENTY-SIX WAS a watershed year—for San Francisco, and for Josephine. Despite the long-awaited completion of a major railroad connecting San Francisco and Los Angeles, the depression experienced by the rest of the country had now hit San Francisco hard, exacerbated by a drought that destroyed grain crops and killed cattle. Unemployment rose precipitously. The gaudy excess of Nob Hill millionaires was stirring up a class resentment that was new to the city by the bay. A general fear settled over the city: perhaps the boom of the prior decades was finally over.

The news of major mineral strikes, and of the sudden fortunes to be made in the mysterious regions of the unsettled frontier, was irresistible. "There was far too much excitement in the air for one to remain long a child," Josephine recalled. There were rumblings of new mining discoveries, sending visions of gold swirling like dust motes in the atmosphere of San Francisco. This could be her destiny, Josephine felt—to take part in this next dramatic round of exploration and building.

Into that atmosphere of anxiety and anticipation came a single spark of rebellion that would reverberate throughout the rest of her life.

THE SPARK CAME from the stars.

From its first performances in London in 1878, *HMS Pinafore* ignited a global craze. Uninhibited by any international copyright protection, Gilbert and Sullivan's words and music flew from coast to coast, and so did people and sets, aided by the rapid advance of rail transport, which enabled a whole production to be relatively mobile. The bold stroke of having the star of the show utter the word *damn*, the "big D" of the script, was an additional guarantor of financial success to the comic opera's local backers, if not rewarding the "damned" originators, Gilbert and Sullivan. Already wildly popular in England, *Pinafore* became an American sensation. *Pinafore* companies sprang up everywhere, matched to every slice of the American demographic: black, white, German, Italian, professional, amateur, children. With over a hundred different opera troupes active in the United States, the marketplace for performers was brisk.

Enter Josephine.

Already seriously stagestruck, she was taking music and dancing lessons. San Francisco had performance schools to suit every pocketbook, and Sophia chose the McCarthy Dancing Academy, operated by Lottie and Nellie McCarthy and their two brothers. With more enthusiasm than talent, Josephine was enrolled as a voice student under the coach Mrs. Hirsch, whose daughter Dora became Josephine's closest friend. Two coquettes who stood just over five feet tall, Dora and Josephine shared confidences and girlish dreams of fame and romance.

The *Pinafore* business was so hot that bookings were sometimes made before actors were hired. Theater companies recruited actors wherever they could, including neighborhood schools such as the McCarthy Dancing Academy. "We're going to Arizona, girl!" was the stage whisper heard whenever Mrs. Hirsch turned her back. Rumor had it that one handsome young man who had been a favorite of Josephine's and Dora's had already joined a production that was being put together by the popular actress Pauline Markham, who had left a starring role in Lydia Thompson's "Blond Beauties" show to start her own traveling Gilbert and Sullivan company with her husband. As was customary, the new troupe would be named after its female star. The stage sets were built, the venues were booked, and the itinerary was set. But at the last minute, the Markham Company lost some of their actors. Perhaps they had not been paid, or perhaps they revolted when they heard about their destination, the Arizona Territory, where Apache attacks were still much feared.

"Have a little more patience," the newspapers pleaded, and reported that fresh faces were on the way to join the troupe.

In vain, Mrs. Hirsch strenuously warned her students, including her daughter, about the empty promises of theatrical producers. But her pleas fell on deaf ears. Dora dreamed of being an actress, and when the Markham troupe began recruiting, she jumped at the chance. "Think of it, Josie, we're going on the stage," she exulted. Josephine caught some of her friend's fever, though in Josephine's case, it was tempered with more than a little chagrin at the pain that she might inflict on her mother. Josephine had confidence in her looks but knew that she was not a great dancer. Perhaps she would be good enough to succeed in the mysterious world of "the Territory."

In the fall of 1878, with the logic of an eighteen-year-old who dreaded her fate as a Polish-Jewish *hausfrau*, Josephine decided to join Dora, a "giddy, stage-struck girl" drawn away on a grand adventure by an even more strong-willed friend.

The first days of the journey were terrifying. Josephine sailed away from San Francisco in tears, overwhelmed with guilt and the desire to return to her mother. She cried even harder when she sacrificed her thick shining braids to play the part of a British sailor boy, a loss that wounded her vanity as well as her self-assurance, though she took some comfort in the novelty of wearing skintight breeches that exposed her trim ankles.

Her fellow actors seemed like beings from another world. It did not help that Dora had thrown herself entirely into the world of the theater. "You can't stay at home all your life and be an actress," she taunted Josephine.

Using the stage names May Bell and Belle Hewitt, Josephine and Dora went through Los Angeles, then San Bernardino, finally drawing close to the Arizona Territory, passing through an alien landscape of desert and mountain passes that reinforced Josephine's sense of estrangement, guilt, and doubt.

THE MARKHAM "PINAFORE on Wheels" troupe performed before capacity crowds throughout the Arizona Territory. In December 1879 they reached Tombstone, a recent addition to the schedule, and still a jumble of miners' tents and a crude facility that passed for a theater. As Josephine walked the dusty streets with the actors and musicians, she might have seen another group of newcomers unpacking their wagons and getting settled: Wyatt Earp, his brothers, and their wives had also arrived in Tombstone.

Markham's troupe was there for a few days and soon back on the road, headed for the next stop on the theatrical circuit. "I remember very little of the details," Josephine would say later, recalling mostly homesickness but also the occasional thrill of being onstage, feeling as if she floated above the floor. Dora had little sympathy, but Pauline Markham took notice of her protégé's confused state and offered the young girl some comfort.

Josephine's tears for her mother soon turned to terror when the actors were intercepted by fierce-looking riders on horseback. Famed Indian scout Al Sieber led one group, and a local lawman named Johnny Behan led the other. Savages on the warpath! they warned. With great gusto, Josephine later embellished the drama of how she and the other Markham actors were saved from unspeakable atrocities. In her narrative, Mexican thieves became renegade Apache warriors. "If those *caballeros* from Sonore get away without being killed or captured it will be a miracle and really too bad," reported the *Weekly Arizona Miner* on December 5, 1879, eager to see Sieber and Behan return with their quarry.

Alas, the posses returned empty-handed, no thieves or savages, but by then, Deputy Sheriff Behan had his eye on another conquest.

Not much taller than Josephine, and about sixteen years older, Johnny Behan was a small, dapper man. He was fastidiously groomed, and she thought he looked great on horseback. In contrast to the tightly knit world of the Jews of San Francisco, where everyone seemed to know everyone else, Behan was a total stranger. She had never met anyone like him before: he was no peddler or baker, but a deputy sheriff with a dashing air of authority. She liked his merry black eyes and roguish smile. But most of all, she liked how he concentrated his whole attention on her, looking past all the other girls.

Older and wiser, Pauline Markham warned her young friend about Behan. He was married and had a child. Josephine assured her that their flirtation was but a "diversion from homesickness."

After some sweet moments whispering with Johnny in the moonlight, Josephine left town with the rest of the troupe, continuing on to Prescott.

And then everything fell apart. Pauline Markham unexpectedly dumped her husband and disbanded the company. On January 28, 1880, the San Francisco newspapers raised eyebrows by publicizing the end of the marriage: "Miss Pauline Markham of the Pinafore Troupe, left last night for parts unknown. Domestic difficulty is supposed to be the cause. She left a note to her husband stating that it was no use to follow her. The direction taken by her is a profound mystery."

Josephine might have chosen to stay in Prescott, but she acknowledged that Johnny had dropped not a word about marriage into his avowals of love. Was he in fact divorced, as he claimed, or was Miss Markham right? Filled with doubt, she contacted her family, and her older sister and brother-in-law dispatched a family friend to bring her home. Although Behan protested that he was a free man, Josephine allowed herself to be taken back to San Francisco. "When you get your divorce," she told him, "come and see me."

THE BRIEF ACTING career of Josephine Sarah Marcus was a secret known only to her immediate family. That first experience in the Arizona Territory set the pattern for the conflicting impulses that would always govern her choices—a hunger for adventure and romance, and a craving for status and respectability. Her failure in this foray from home was a source of humiliation and regret.

Josephine attempted to draw a complete veil of secrecy over this part of her life: "Except for the bearing my escapade had upon what followed in my life and to quiet all erroneous rumors with the truth I should not have felt constrained to relate it." In fact, she denied that she had ever been on the stage. "I have heard it whispered that before my marriage I was a dancer," she stated with an excess of dignity. "The rumors placed my dancing career all the way from the Bird Cage Theatre in Tombstone, Arizona, to a dance hall in Nome." Later, she would lead her biographers on a wild goose chase that had her abducted by Native Americans, instead of running away by her own choice.

This first chapter of her adventure was over, and Josephine was back in San Francisco in time to be counted for the 1880 census as a member of her parents' household, which was even more crowded now that the Marcus family included her half sister Rebecca's husband and four children. Little else had changed. Josephine had come home with nothing to show for her adventure other than a mild but inconvenient case of what was probably St. Vitus's dance—a neurological illness that stemmed from a childhood case of rheumatic fever. Josephine blamed her lassitude and sudden lack of coordination on exhaustion and the anxiety of life on tour. Lacking modern penicillin, the common treatment in 1880 would have been a quiet convalescence at home. But Josephine did not take well to the restraints placed on her activities. After the excitement of Arizona, her medical confinement seemed like house arrest. She was pining for Johnny, and chafing at the sense of freedom interrupted.

True to her last command, Johnny did come to San Francisco to plead his case. Undeterred by her departure from Arizona or her return to her Jewish family, he met Hyman and Sophia Marcus and declared his intention to marry their daughter, presenting Josephine with a diamond ring. Although no account remains of the Marcuses' meeting with Johnny, it is not hard to imagine their consternation at Josephine's odd choice. Johnny's religion was the least of his strangeness to Mr. and Mrs. Marcus. Josephine herself vacillated between infatuation and doubt. She sent Johnny away with no promises for the future.

But Johnny did not give up easily. He sent an emissary from Tombstone to intercede on his behalf, thinking perhaps that the respectable wife of a successful attorney would overcome Josephine's reluctance.

He chose his advocate well. Ida Jones (also known as Kitty or Katherine) came to the Arizona Territory with her husband, Harry Jones, a respected lawyer. The difference in age between Kitty and her husband was about the same as between Josephine and Johnny. Behan was indeed an unmarried man, Kitty reassured Josephine, and he loved her most sincerely. Kitty painted a rosy picture of Josephine's future life, how she could tie her fortunes to Tombstone's rapid growth and Johnny's excellent prospects.

It was time for Josephine Marcus to give the frontier another chance. Her misgivings melted away under the pressure of her own boredom and restlessness, as well as Kitty's persuasive arguments. She dismissed her parents' concerns, and they were powerless to stop her. Josephine was in the grip of "an enticement that would beckon to her all her life and that she could never resist—that was the lure of the prospect trail."

By December 1880, Josephine was back on the dusty road to Tombstone.

Kitty Jones had replaced Dora as Josephine's companion now, completing the last part of her assignment to deliver the reluctant bride into Johnny Behan's arms. Josephine felt comfortable with her new friend's lively southern ways and cheerful disposition. The two women were joined by Mariette (also known as Maria) Duarte, a demure Mexican senorita who, like Josephine, was on her way to meet her future husband. The three women munched on gumdrops, while Kitty kept up a gay stream of chatter to allay the nervousness of the two younger women.

They traveled by stagecoach from Benson to Tombstone. Just before departure, a handsome young man climbed up and took the seat next to the driver, tipping his hat to Mrs. Jones. Kitty identified him as Morgan Earp, shotgun messenger for the Wells Fargo Company and one of the Earp brothers. You will be meeting the rest of those tall, blond Earps in Tombstone, she assured Josephine.

As they entered the town, Josephine was amazed by the dramatic changes it had undergone. Last year's tent city had become the largest city in the Arizona Territory, more than twice the size of Tucson or Phoenix. In place of the "overpowering silence" she had experienced on her last visit, the desert air was filled with the smell of leather and horses, a cacophony of construction and commercial activity, and the constant din of hammers mingled with the shouts of business and social exchange. The mines never stopped. Large ore wagons drawn by mules rumbled along crude streets that had not existed the year before. These were the sounds and sights of a town being created at a frantic pace, by settlers determined to take advantage of an opportunity that might only appear once in a lifetime.

The area now known as Tombstone was discovered by an enterprising miner named Ed Schieffelin, who persisted in his search for silver despite the remote and forbidding landscape. Soldiers at a nearby fort predicted Schieffelin would die of thirst or in an Apache raid, and warned him that he would find not his fortune but his tombstone. "The remarks being made often impressed the [word] on my mind," he recalled. After finding an unusually rich lode of silver, he organized the first mining district in April 5, 1878, and had the droll idea to name his first claim "the Tombstone," then the next ones "Graveyard No. 1" and "Graveyard No. 2."

Tombstone would become the most famous boomtown in Arizona. Its first mayor, John Clum, described the town as having been conjured by the "greatest of magicians" and marveled that an incredible city could spring up from a vast empty landscape of boulders, mesquite, and cactus. At first, the residences and stores were rudimentary and small. Almost everything was made of wood, which made the Fourth of July fireworks terrifying for those who imagined the rockets raining down like giant matchsticks.

A rush of capital from New York, Philadelphia, and Boston assured that Tombstone would grow up in a hurry. The all-important Wells Fargo company opened an office, a sign that the boomtown had crossed the threshold of commercial viability, big and stable enough to warrant attention from the ubiquitous business agent of the West, "the ready companion of civilization, the universal friend and agent of the miner, his errand man, his banker, his post office." One hundred new people arrived by stagecoach each week, not only miners and bankers, but engineers, lawyers, and merchants. The next wave brought their wives and prostitutes, with a rush of French restaurants and oyster bars, tennis courts and bowling alleys, springing up to serve and entertain them.

Despite the dust and the summer heat, Tombstone had style. Several dress shops and milliners outfitted the local women with the latest Paris fashions. Tombstone had "the most expensive of everything," proclaimed an Arizona travel guide in 1881. Two luxury hotels, the Cosmopolitan Hotel and the Grand Hotel, competed for customers with bars and casinos and public rooms outfitted with chandeliers, carpets, grand pianos, and paintings, assuring a level of elegance that would not be out of place in San Francisco.

When Josephine climbed down from the Benson stagecoach, Kitty's husband Harry and Johnny Behan were waiting for her. Maria Duarte disappeared with the rest of the passengers into the Tombstone crowd.

Harry Jones had an office on Allen Street, right in the center of Tombstone, and a comfortable three-room residence across the street from Johnny Behan. But Josephine did not accept the Joneses' offer of a place to stay. Instead, she walked across the street to Johnny's house. There she discovered that he had a son who would be sharing her new home.

Josephine's initial hesitation was correct; Johnny Behan was escaping a failed marriage and a stalled career. He had married the former Victoria Zaff in San Francisco in 1869 and settled in Prescott, near her parents, fathering two children during their six years of marriage. It was true, as Johnny boasted, that he'd held elected and appointed offices, including stints as sheriff and as an elected member of the territorial legislature from Yavapai County. But he had lost his last election, and then his marriage and his political career began to fall apart.

Victoria filed for divorce in 1875, going enthusiastically public with graphic allegations of Johnny's violence and debauchery. She charged that her husband was often drunk and subjected her to "a threatening and menacing manner calling me names such as whore and other epithets of like character and by falsely charging me with having had criminal intercourse with other men, threatened to turn me out of the house, quarreling with, and abusing me, swearing and threatening to inflict upon me personal violence." Johnny frequented the local brothels, causing a scandal among their friends and neighbors, one of whom testified in support of Victoria's accusations, though dryly noting, "I can't say of my own knowledge that [Johnny] had carnal intercourse with the inmates of said houses of ill fame."

"I have been nearly driven to distraction," testified Victoria Behan, insulted and disgusted by Johnny's outbursts and his attachment to one particular house of ill fame and one favorite prostitute, Sadie Mansfield, with whom Johnny did "carouse, cohabit, and have sexual intercourse."

Behan denied everything. But the damage was done. Victoria had incited Prescott and its environs, especially given her stepfather's prominence as the former sheriff. She was granted custody, and Johnny was ordered to contribute child support for their son Albert (their daughter died soon after the divorce). However, Victoria voluntarily allowed Johnny to take Albert to live in Tombstone, probably because she was already planning her second wedding, this time to a local businessman.

Behan and Albert, accompanied by Johnny's favorite racehorse, Little Mare, moved to Tombstone, where Johnny planned to resume his career as lawman, businessman, and politician. A new county would be created out of the bustling Tombstone region, and he hoped to be appointed as its first county sheriff. He invested in a livery stable with his well-connected friend John O. Dunbar, and they also leased the bar at the Grand Hotel.

Josephine would build two lasting relationships in Tombstone. One of them was with Johnny's son, Albert, who immediately warmed up to the pretty young woman who had moved in with his father. The other lasting relationship would not be with Albert's father.

Johnny introduced Josephine around town as Mrs. Behan, but Josephine was quickly disenchanted with the empty promise of the salutation. Johnny's conduct was inexplicable. He refused to set a wedding date and avoided any reference to marriage. In her letters home, she hinted enough of her unhappiness to prompt her father to wire her money.

What happened next was predictable, if sad. Josephine taunted Johnny with her financial independence, which prompted Johnny to renew his marriage proposal, take Josephine's parents' money and her diamond engagement ring, and sign a ninety-nine-year lease on a nice home on Safford and Sixth—listed solely in his name.

Johnny Behan had successfully restarted his life. He had acquired a desirable new mistress who acted as governess to his son, a responsibility that she gladly accepted out of affection for the young boy. Johnny's professional life was flourishing: the Grand Hotel bar business was thriving, he was well liked in Tombstone, and he balanced his Democratic connections with a close business partnership and friendship with the Republican Dunbar family. He had correctly anticipated the creation of the newly named Cochise County and was massaging the political connections that would smooth his path to becoming its first sheriff.

Wyatt Earp was the only hurdle in his way.

THE EARP WAGON came rolling into Tombstone in December 1879, laden down with furniture and pots and pans, a sewing machine hanging off the back. The family looked more like a bedraggled pack of peddlers than smooth gamblers and cool lawmen. Their timing was excellent. Although founder Ed Schieffelin had already cashed out and would soon be headed for Alaska in search of the next big strike, Tombstone had not yet reached the pinnacle of boomtown prosperity.

Even to their wives, Wyatt and his brothers were like "peas in a pod." Virgil was the eldest, and he had a certain twinkle in the eyes that his brothers, especially dour Wyatt, rarely betrayed. Josephine would later agree that Virgil's face was "filled with laughter and he was more free with his mannerisms than the others." To the townspeople, however, the Earp brothers looked so much alike that one merchant sold a horse to Morgan and swore that he sold the horse to Wyatt. All of the brothers were light-haired, wore bushy mustaches, and topped six feet. When they walked down the street together, they were an imposing group, with lean, rugged physiques and a certain athletic grace.

Among the Earp brothers, Virgil was the most responsible and Wyatt the most handsome. One of the earliest Tombstone chroniclers, Walter Noble Burns, captured Wyatt as "the lion of Tombstone":

> *His face was long and pale, his deep set eyes were blue gray, chin was massive, heavy tawny moustache, hair as yellow as a lion's mane, deep voice was a booming lion-like growl, and he suggested a lion in the slow, slithery ease of his movements and in his gaunt, heavy-boned, loose-limbed, powerful frame.*

Burns had never seen young Earp, but Burns's flowery prose was matched by contemporaries who knew Wyatt in his prime. When Kansas judge Charles Hatton was interviewed about Wyatt, his wife interrupted to add her two cents that he was "the handsomest, best-mannered young man in Wichita." Men were no less effusive and lingering in their description of Earp's hunky good looks. Wyatt's close friend Bat Masterson described Earp as "weighing in the neighborhood of one hundred and sixty pounds, all of it muscle. He stood six feet in height, with light blue eyes, and a complexion bordering on the blond." John Clum, the mayor of Tombstone, recalled Earp as "tall, erect, manly, serene, and in neat attire . . . I still have a clear vision of that dignified figure walking calmly along Allen Street." Clum went on to praise Wyatt's manner as friendly but reserved, "equally unperturbed whether he was anticipating a meeting with a friend or a foe."

The Earp brothers preferred each other's company to that of any outsider. All veterans of an outdoor life and a frontier philosophy that had toughened them, they relied on each other. Their wives were equally clannish and mutually supportive. But Tombstone would be the first time—and the last—that all of them lived and worked together.

THE EARPS BELONGED to an old Scotch-Irish family who immigrated to Maryland around 1680. Ever a roving clan, generations of Earps changed their residences as the borders of the young country expanded, crisscrossing the thirteen original colonies and eventually wandering westward to the frontier.

Wyatt Berry Stapp Earp was born on March 19, 1848, in Illinois, the fifth of the eight children of Nicholas and Virginia Cooksey Earp. Named for his father's commanding officer in the Mexican War, Wyatt was too young to serve in the Civil War, but three brothers fought on the Union side. Oldest brother Newton was unharmed, but James Earp was injured enough to qualify for a lifetime pension. Virgil suffered a less visible wound: before going off to war, he had married without the consent of his bride's parents, who took advantage of his absence to spirit the young woman away. They told Virgil that she was dead. It would be nearly thirty years before Virgil would discover what had become of his young wife.

As the Civil War drew to a close, the Earp family uprooted themselves once again to join one of the great migrations of American history, the fulfillment of the "manifest destiny" that had driven American policies and demography since the founding of the country. Nicholas Earp was asked to lead a wagon train headed to California. A strict disciplinarian with a salty tongue, Nicholas soon alienated some of his constituency with his profane language and his intolerance of noisy and active children, whom he threatened to whip if their parents did not. Tensions were high during this long and dangerous journey. As attacks by Indians were relatively rare, deprivation and disease caused the most fatalities during the seven-month journey, especially for children and the many babies who were born along the way. While men emphasized the adventure and the economic objectives of the trip, the women were consumed with the difficult communal life of the wagon train, especially the challenges of everyday tasks like child care, laundry, food preparation, caring for the sick, and hygiene. It was the women who counted the horrifying number of roadside graves and recorded the terrors of the journey in their journals and letters.

By the time Nicholas Earp's wagon train had arrived in San Bernardino, California, in December 1864, the accompanying families were grateful to have reached their destination—and eager to say good-bye to the demanding Nicholas Earp.

Despite the ardors of the trip, the Earp family stayed on the West Coast only a few years before returning to the Midwest. They took up residence in Iowa, and then in Lamar, Missouri. At twenty-two, Wyatt ran for election as the town constable, and fell in love with the local hotelier's daughter. He married Aurilla Sutherland on January 10, 1870, with Nicholas Earp presiding as justice of the peace. Of the four women who would eventually take the name Mrs. Earp, only Aurilla would celebrate with Wyatt in a public ceremony.

With a legally sanctioned marriage, a responsible job, and most of his family gathered around him, Wyatt bought a house and showed every sign of being ready to settle down and become a pillar of the community.

Less than one year later, Aurilla died in childbirth during a cholera epidemic; her son was buried with her. Wyatt never spoke publicly about his young bride, other than to acknowledge that he had been married. He never admitted that his wife's death set off a personal crisis as well as a family feud.

The next few years would prove to be the darkest of Wyatt's life. The grieving young widower brawled in public with Aurilla's brothers, stole horses, and served a month in jail before skipping town. By the time his official biography was published, it would not mention any of his times in "the cold and silent calaboose" as a gambler, pimp, and thief. Josephine may have known only the most general facts about these years. Without acknowledging the aftermath of his brief marriage, she shaped the timeline around his years on the plains, where an unusual fraternity of some ten thousand buffalo hunters honed their drinking and shooting and gambling skills and spawned lifelong friendships. It was here that Wyatt met Bat Masterson, whose career would parallel Wyatt's, and who would become Josephine's friend and admirer. However, during these years, Wyatt was at least as familiar with brothels as he was with buffalo and Bat Masterson.

In cities like Peoria, Illinois, and Wichita, Kansas, prostitution was publicly frowned upon but privately tolerated and regulated as an important source of revenue. It was common practice for mayors and police officials to impose monthly fines or jail terms, castigating the offenders with loud public outcries. A few days later, these revenues safely stored in the city coffers, business would go back to normal. The city treasuries relied on the prostitutes' contribution, and local merchants depended on the "sporting houses" to keep visitors happy and the town lively.

Wyatt was arrested at least twice in Peoria before leaving for Kansas, where he ran brothels with his brother James and James's common-law wife Bessie. In Wichita he entered into a second relationship, though this one had no officiating justice of the peace. He lived for more than a year with Sally Haspel, who was probably the daughter of a local madam who worked with Bessie. Newspaper accounts identify "Sarah Earp" as depraved but good-looking. Both Bessie and Sally endured a grueling cycle of arrests and fines until the spring of 1875, when Wyatt joined the local police force. At this point, the monthly payments ceased, and Sally disappeared from the public record.

Wichita and Dodge City were major hubs in the Kansas cattle trade and temporary homes to throngs of young men returning from months on the open range. "Cow-boys" meant cash to the local businesses, but signaled trouble. When they came to town, soaked with alcohol and tense with gambling challenges, their antics destabilized the fragile peace of a gun-toting community.

Wyatt served in Kansas with distinction as a peace officer. He was paid in cash for each arrest, and mostly he arrested cowboys. Town leaders liked him because he managed to avoid fatalities; most unruly cowboys responded well to a sharp knock on the head, known as buffaloing, and an uncomfortable night in jail. This technique worked nicely because the troublemakers were mostly transients blowing off steam. Things would be different in Tombstone.

Wyatt's world of gambling and guns and rowdies brought him into contact with a rising Dodge City gambler, Doc Holliday, a thin, tubercular dentist from Georgia. Their friendship was sealed when Doc Holliday came to Wyatt's aid in a shootout, tossing a gun to him at an opportune moment.

Loyalty to friends and family was an absolute for the Earps. "Doc was my friend and I was Doc's friend until he died," Wyatt steadfastly declared, although none of his brothers and few of his friends shared Wyatt's affection for Doc, whose nasty temper was as legendary as his dangerous straight shooting. Bat Masterson detested him. "Physically, Doc Holliday was a weakling who couldn't have whipped a healthy fifteen-year-old boy," Masterson recalled, dismissing Doc as high-strung and effeminate.

Even with Doc around for excitement, Dodge City was, in Wyatt's words, "losing its snap." Virgil was scouting the latest boomtown, and urging his brothers to convene there. In September 1879 the Earp caravan set out for Tombstone. Among the party were James and Bessie and her teenage daughter. Doc and his longtime companion and lover, "Big Nose" Kate, had made plans to join them later. Younger brother Morgan Earp— everyone's favorite—and his common-law wife Louisa were also on the way. And there was a new Mrs. Earp: in place of Sally, Celia "Mattie" Blaylock was now sitting next to Wyatt. She had left home as a young girl and was living on her own until she met Wyatt, probably in Fort Griffin, Texas. No beauty, Mattie was a sturdy, large-boned woman with a sweet square face, a mass of curls hiding a broad brow, and long ringlets down her back. She had the look of someone dependable, shy, and long-suffering.

Although Nicholas and Virginia Earp had been joined in a traditional marriage, their sons preferred common-law partnerships requiring nothing other than a voluntary declaration that man and woman were living together as a married couple. This was a popular alternative in the western territories, where couples moved between remote and sometimes dangerous locations where clergymen were not always available. "If a man and woman said, 'We are Mister and Missus,' they were, that is all there was to it," recalled one observer of frontier marriage. The self-declared husband and wife could sign legal documents together and were classified as married by the census, but the common-law widow was not eligible to collect her disabled husband's Civil War pension. And, as Mattie Earp would discover, common-law marriages were easy to create, and could be dissolved just as easily.

In Prescott, Virgil and his common-law wife Alvira Packingham Sullivan Earp (known to all as "Allie") joined the caravan. They had been living in the town where Virgil had been appointed deputy U.S. marshal, the highest office that any of the Earp brothers would hold. Virgil and Alvira had been a couple since 1874, when Allie served Virgil a meal in a Council Bluffs, Iowa, restaurant. It was love at first sight: Allie was immediately drawn to this "big blond" who "looked nice on a horse." Small and spare, Allie adored Virgil and was the most quick-witted of the group, always ready with a sarcastic retort for anyone other than Virgil.

The last Earp to arrive in Tombstone was Morgan. He too had acquired a common-law wife, Louisa Houston. A longtime sufferer from rheumatoid arthritis, Louisa stayed behind with Morgan's parents in San Bernardino, California. "They are all old fashioned people here," she confided to her sister, "and I like it very much." Despite Nicholas's harsh reputation, he and his wife treated Louisa with exceptional kindness, and her health improved in the temperate climate. She knew from Morgan's letters that Tombstone's steady rains and dust storms did not augur well for her health, but she missed him and left San Bernardino as soon as he sent for her.

The Earps came to Tombstone to make money. Everyone dreamed of striking it rich, but Wyatt and his brothers knew better than to rely solely on the vagaries of mining. Wyatt's initial plan was to start a stagecoach line, but he discovered belatedly that the town was already well served. He was not discouraged. "There's other business here, plenty of it," he anticipated, and with the help of his solid recommendations from Wichita and Dodge City, he signed on as a shotgun messenger for Wells Fargo, while investing with his brothers in mines, water rights, and gambling concessions. But he still aspired to a position on the law force, like Virgil, who had been named chief of police. In a town like Tombstone, peacekeeping went along with lucrative tax collecting.

And Tombstone certainly needed lawmen.

When the Earps joined the community of Tombstone, they encountered the unique social structure of a well-developed boomtown. What looked superficially like a wide-open frontier town was nearly as stratified as Josephine's San Francisco, albeit along different fault lines. The Earp men mixed among political and business leaders, but they did not socialize with them. They might have a drink with the mayor or the biggest mine owner in town, but they were not invited over for supper. Their common-law wives were shunned, not only because they were poorly educated and living in unions unblessed by judges or clergy but because their husbands were gamblers and saloonkeepers. All legal pursuits, to be sure, and considered essential in a frontier town, but not embraced by polite society.

Neither Josephine nor the Earps participated in the conventional world of church socials, literary nights, amateur dramatic benefits, and lending libraries, though these were happening all around them. That Tombstone can best be seen through the eyes of a trio of firsthand observers, whose contemporary diaries and letters offer a compelling picture of Tombstone's aristocracy: Clara Spalding Brown, George Parsons, and Endicott Peabody.

Clara Spalding Brown came to Tombstone with her husband a few months before Josephine, enduring a stagecoach trip so bad that "nothing short of a life and death matter" would tempt her to travel again during the summer. A talented writer and loyal San Diegan, she began her distinguished journalism career with a series of colorful, keenly observed letters to the *San Diego Union*. She drew social distinctions with a fine and knowing hand: "There are frequent dances, which I have heard called 'respectable,' " she relayed to her readers; "but as long as so many members of the demi-monde, who are very numerous and very showy here, patronize them, many honest women will hesitate to attend." Those honest women preferred more exclusive balls with good musicians and refreshments served by "darkies, in imitation of metropolitan style."

George Parsons came to Tombstone early in 1880. The eldest son of an affluent and well-educated eastern family of lawyers and bankers, Parsons had been a bank teller in San Francisco and came to Tombstone to chase his dream of striking it rich. He was warmly welcomed into the homes of the mining magnates who were the gentry of the Arizona Territory and the first families of Tombstone. Unlike Brown, Parsons wrote in private about his daily life, mostly short, telegraphic journal entries that registered the weather and political climate and catalogued his business dealings, as well as his social engagements and random observations that interested him, such as the atmosphere of murder and mayhem in Tombstone, or the sight of a lesbian couple embracing in public.

The third member of this unusual Greek chorus was Endicott Peabody, the future founder and headmaster of Groton, where he would one day teach the school's most famous graduate, Franklin Delano Roosevelt. As a handsome young newly ordained minister, Endicott enjoyed his cigar and his drink, and riding enthusiastically around Tombstone and its environs, recruiting congregants. A frequent letter writer, he chronicled life in Tombstone for the enjoyment of his friends and family back east, expressing his surprise at the sophistication of Tombstone. Time would tell if he would stand by his early observation: "The gamblers said to be a very decent lot, Cowboys alone bad."

The Earp wives lived only a short distance from Brown, Parsons, and Peabody, but they might as well have been a million miles away. They cleaned their houses, socialized among themselves, and made money with Allie's prize possession, the sewing machine that they used to mend canvas tents for the miners for a penny a yard. Other than Bessie, none of them had children, perhaps in deliberate recognition that babies and boomtowns were not well matched. The Earp women knew nothing of cowboys or politics. They hardly ever visited "downtown," though it was five minutes away. On one occasion Allie and Mattie ventured out and returned home tipsy, almost too terrified to face their husbands. As Allie recalled:

> *Me and Mattie, Wyatt's wife, wanted to go down and peek in the nice hotels and restaurants. So on a terrible hot morning when the men was away we went and had a good time lookin'. Then we met a friend who gave us a sip of all different kind of wines, some real fine. We got home and in bed all right, and everything would have been jim-dandy but Wyatt and Virgil came home for dinner for the first time during that hot spell. All I remember is waking up and seeing Virgil sittin' by the bed stiff as a poker and Mattie spillin' the coffee Wyatt was makin' her drink.*

"That was our life: workin' and sitting home. Good women didn't go any place," Allie remembered. "Everything was nice if you had money, and we didn't so it wasn't."

For drifters like Allie and Mattie, attachment to their husbands and extended family was everything. Loyalty was unconditional, and the Earp brothers were beyond anyone's criticism, though in private, Allie struggled to find anything to praise in Wyatt. She found him serious, self-centered, and intimidating, like Nicholas Earp. Where Virgil was always ready with a joke and a loving word for his wife, Wyatt mocked Mattie's passivity and may have also complained about her fondness for laudanum, an opium-based painkiller that was a common remedy for toothache or cough or insomnia.

JOSEPHINE MARCUS BEHAN'S social status was even murkier than that of the Earp wives. She was caught in a netherworld between wife and mistress, stepmother and governess. Johnny continued to escort her to social affairs, where she mixed with couples who were legally married, such as the Dunbars and the Joneses. After all, she argued to Johnny, couples did marry, even in Tombstone. For example, Maria Duarte, her stagecoach companion, had become the lawfully wedded Mrs. Pete Spence. Even James Earp's stepdaughter had married a local businessman in a legal ceremony, despite the fact that her mother, former madam Bessie Earp, was in a common-law relationship, as were the rest of the Earp women.

Josephine had the additional complication of being Jewish, which she never hid but never used to bridge her social isolation. Tombstone had an active Hebrew Association, and counted a considerable number of Jews among its mining executives and merchants. German or Polish heritage mattered far less in the frontier, so she was no longer subject to that prejudice. She had no interest in socializing within the Jewish families or meeting any of Tombstone's eligible Jewish entrepreneurs, but gravitated toward those who played a distinctive role in the frontier—the lawmen and the gamblers. She was friendly with Sol Israel, the Jewish proprietor of the Union News Depot, but in a town where newspapers were extremely popular, everybody knew Sol.

Josephine's once clear vision of her future as Johnny's wife was fading. As she walked through the streets of Tombstone, conducting the ordinary business of picking up her mail, buying groceries, or ordering a new dress, she was known to all as Mrs. Behan. She received her mail under that name, and may have told her parents that they were already married. Her family was far away; although there were several Marcuses listed in the hotel registers of the time, they were most likely traveling merchants, not Josephine's relatives.

IN 1881 TOMBSTONE was three years old—and for those keeping score, it was approaching the upper limits of an average boomtown lifespan. True to the timetable, and despite the frantic activity everywhere, the easy days were over. Tombstone's biggest silver deposits had been mined. Copper was the new thing; the boom in nearby Bisbee was already siphoning off people and capital investment. California senator George Hearst hired Wyatt Earp to accompany him on a tour of the Tombstone mines; they became friends, but Hearst declined any Tombstone investment.

Beneath the glittering surface was a city growing carelessly, built on top of a shaky foundation. Disputes over mining and real estate claims kept the city's lawyers busy. The mines were experiencing water seepage. Sanitation was rudimentary, and fire was a constant concern. Because of the scarcity of water, local editorials warned that the town was a "tinder box, and liable any day to be swept from existence." Gas lamps were a recent innovation, but they merely illuminated the filthy streets and the rats. Cowboys were running wild, and not the relatively harmless variety known to Wyatt in Dodge City, young cowhands who rode into town after months on the trail, looking to have a good time, make a lot of noise, and return to the range. A more dangerous lot, Tombstone cowboys had friends in high places. Clara Brown defined cowboys as "a convenient term for villains," and George Parsons considered them a synonym for "rustler" and "desperado—bandit, outlaw, and horse thief."

Crime was on the rise. A transaction at a Tombstone bank was a "special event" that called for "special precautions," noted Endicott Peabody wryly. When entering the bank, "one came to the receiving teller with a pistol near to his hand. The paying teller was still more fortified, while, at the back, on the manager's desk, lay a pistol; at his side was a gun, and in a box at the other side of the barrier at which a customer of his would be standing was another pistol which, unbeknownst to him, would be pointed directly at his diaphragm."

As concern about crime rose, the question of who would be the first sheriff of Cochise County became more pressing. Josephine assumed that the decision was important to her future, since the county sheriff kept the peace but also collected the taxes and the fines. The city fathers passed ordinances against gambling, prostitution, and carrying of firearms, but they were less concerned with the thankless task of regulating public morality in a frontier town than with keeping the peace and raising money for the town's essential services—including paying the sheriff.

If it was money that would finally drag Johnny to the altar, then Josephine wanted him to become sheriff of Cochise County.

Johnny Behan and Wyatt Earp openly campaigned for the position. The two men could hardly have been more different. Johnny was a natural politician: he was "Glad Hand" to Wyatt's "Trigger Finger," as writer Walter Noble Burns would later characterize their rivalry. "Johnny Behan was friendly, Wyatt Earp was grim; Johnny Behan smiled, Wyatt Earp shot from the hip." Dark-eyed, short, and round-faced, Johnny was a lifelong Democrat. Wyatt towered over him, tall and athletic, fair and blue-eyed, and staunchly Republican. Johnny represented the interests of ranchers and was a good friend to the cowboys, while Wyatt stood with the miners and town officials and was a close friend of many of Tombstone's business leaders. As two men-about-town, they competed in horse races and in shotgun competitions. They enjoyed the company of women, though Wyatt tended to be serially monogamous, while Johnny Behan had a soft word and a welcoming smile for all desirable women.

Eager to avoid a public battle, Johnny Behan offered Wyatt a deal: if he would withdraw from the race, Behan would appoint him as deputy. Wyatt obligingly followed the script, but when Governor John C. Fremont appointed Johnny to the position, the new Sheriff Behan reneged on his offer to deputize Wyatt. Johnny's double-dealing made permanent enemies of all the Earp brothers.

Johnny had friends everywhere, and to his detractors, he was far too cozy with the cowboys and rustlers, who resented any restriction on how they lived, how they made a living, and how they entertained themselves. With the population exploding, demand for beef had grown to the point where it was second only to silver as local currency. Cattle thievery was rampant, and among those suspected of being involved in the practice were families like the Clantons and McLaurys, substantial property owners and close associates of Sheriff Behan. As long as Johnny Behan could keep a lid on things, however, and there was an ample supply of beef, the town officials would tolerate some level of cattle thievery.

In the summer of 1881, things spiraled out of control for Johnny. He was struggling to control violence in the streets in Tombstone, while another explosive situation was building at home. He could no longer pretend to Josephine that he was delaying their marriage until his financial situation became stronger: he was now the sheriff, and making plenty of money. Yet the higher his status rose, the less interest he showed in her.

Josephine was still calling herself Mrs. Johnny Behan, but the title had long turned sour. Eighteen eighty-one was not destined to be an auspicious year for Mrs. Behan, or for the city of Tombstone.

THAT SUMMER WAS unbearably hot. With stifling days that Clara Brown called "wringoutable," the city was buffeted by dust storms that swirled through it like a gritty fog, and endured ferocious nighttime thunderstorms that delivered no rain. On June 22 the town was struck by a major fire that began in a barrel of whiskey and then roared through the main part of town, consuming four square blocks, destroying dozens of stores and restaurants, and causing major damage to the Cosmopolitan and Grand Hotels. Firefighting resources were woefully inadequate, and ironically, Mayor Clum was away pricing new equipment. Among the injured volunteer firefighters was George Parsons, who was caught under a collapsing roof and had a nasty encounter with a stick of wood that pierced the left side of his face. Clara Brown's bank was destroyed, a financial loss from which many depositors (including Clara and her husband) never fully recovered.

The town immediately began to rebuild, using adobe rather than wood whenever possible and raising money for fire equipment and an emergency supply of water. The newly organized "Rescue Hook and Ladder Company" was inaugurated with a grand ball to raise money, but attendance was disappointing; most women had already left town to avoid the summer heat. "Now is the time when the fashionable dame packeth her Saratoga and departheth for some haven in the East or West outside the precincts of Arizona," Clara Brown explained. The only women left in Tombstone were what she called the SAH or the "stay at homes," who would have loved to escape Tombstone's baking temperatures with a vacation in the mountains, but were deterred by illness or fears of an Indian attack.

When the rains finally came, floods overran the badly designed Tombstone streets. "We're in a bad way in town," George Parsons wrote in his diary on August 25. "Eggs and potatoes gone and flour getting scarce owing to the wash-outs. No mail at all and things generally are in a deplorable condition."

Josephine was not affected directly by the fire, and any distraction from her personal woes was fleeting. She had no money and no wedding ring. She could no longer tell herself that Johnny Behan intended to marry her. She left their home briefly, possibly to accompany Albert to consult with a doctor, but this good deed met with no reward. She returned home to discover Johnny Behan with another woman.

She had been slow to acknowledge the reality behind Johnny's reputation as a womanizer, so indiscriminate that one of his drinking buddies joked that he'd seen a horse that reminded him of Johnny, always trying to get at the mares. There may have been several lovers and prostitutes competing with Josephine that summer, but one likely candidate was Emma Dunbar, wife of Johnny's business partner and friend, John Dunbar, who either did not know about their dalliance or simply didn't care, since the couple remained together. Emma stayed in touch with Johnny Behan for years afterward, writing to him with a playful intimacy long after they'd both left Tombstone.

Josephine was not inclined to be so forgiving. Whether she confronted Johnny about another woman or discovered the disturbing signs that

Johnny had contracted syphilis, her situation had become dire. It was time for her to leave Johnny. Her next steps were less obvious: her parents would be appalled if she returned home for a second time without a husband. But if she wanted to stay in Tombstone, she would have to find some means of support. Although she was reasonably well educated for a woman of her time, she had no profession or independent source of income.

Models for financially independent women were rare in Josephine's experience. She hadn't known any self-supporting Jewish women in San Francisco. On the frontier, things were different. She had seen for herself the example of Pauline Markham, who made a comfortable living as an actress and theatrical producer. Tombstone had a number of successful women. Josephine undoubtedly knew about Nellie Cashman, who built a string of frontier businesses, including a restaurant and grocery store in Tombstone. Josephine's friend Addie Borland ran a millinery shop. Clara Spalding Brown was a freelance writer. But Josephine was only too aware of her limitations. She was no actress or writer, and had more interest in wearing fine clothes and dining in restaurants than in running a retail establishment.

The most common occupation for a woman in Tombstone was prostitute or performer. Or both: many of the more attractive prostitutes also performed at the theaters and dance halls in town. Tombstone's sex trade was regulated by the city, with license fees and fines that went into the town treasury. Prices were set according to age, ethnicity, and privacy. At the bottom of the ladder were the women who worked alone in the "cribs," crude little shacks with room for a bed and chair. Rates ranged from twenty-five cents for Chinese, black, and Indian women to fifty cents for Mexican women, with higher rates for French and American women. In madam-operated parlor houses, the women charged $15, with a premium for overnight stays or special requests. The most popular prostitutes could make up to $150 per week in private rooms lavishly appointed with fancy furniture and drapery. However, even for the high-priced women carrying engraved calling cards with French names, there were high rates of alcoholism, disease, violence, and suicide. Many of the women were addicted to laudanum. Prostitutes were subject to weekly medical inspections by Dr. George Goodfellow, an Ohio-trained surgeon who treated the women with compassion and skill. His real specialty was gunshot wounds, but he was also the local abortionist.

Although Tombstone residents were well aware of where the prostitutes lived and worked, public solicitation was forbidden. Care was taken to protect married women and children through an unwritten law that kept the "soiled doves" away from the main streets. "Where is the mining camp without its gamblers and sharpers, its courtesans and adventurers?" theorized Clara Brown, who shrugged at the presence of saloons and brothels as an inevitable feature of boomtown life, though she did not expect to set a place for any of these men and women at her table.

Josephine worked tirelessly to obscure this troubled time of her life, and she was mostly successful. She left Johnny, but with the exception of a few money order receipts, Josephine Marcus or Josephine Behan disappeared from the public record for about six months. Her residence was unrecorded. She may have kicked her faithless lover out of the house they shared; it had, after all, been purchased with her parents' money. Out of affection for young Albert Behan that would last a lifetime, hesitant to leave him with his faithless father, she may have stayed in that house with the child while she took some time to plan. Or she may have moved into one of the local rooming houses or found shelter with Kitty and Harry Jones, or with John and Annie Lewellen, a family who lived nearby.

To add to the mystery of where and how she lived, Johnny's favorite prostitute Sadie Mansfield, the particular focus of his wife's fury in their divorce suit, had followed him from Prescott to Tombstone, where she became known as "Forty-Dollar Sadie." Decades later, this would lead to speculation that Sadie Mansfield and Josephine "Sadie" Marcus were one and the same. They were not, according to Tombstone old-timers who knew both women. Moreover, given Josephine's pride and the option of appealing to her parents, it is unlikely that Josephine would have risked even a temporary stint as a prostitute.

Josephine conserved her meager store of capital and pondered her next step. As one of Clara Brown's so-called SAH women, she was a prisoner of Tombstone. She could find a job. Or she could find a new lover. There were nine men for every woman in Tombstone.

JOSEPHINE MET WYATT in the summer of 1881, most likely at Sol Israel's Union News Depot on Fourth Street.

They had been circling each other for the better part of a year. Unlike the Earp wives, Josephine did not keep herself hidden at home but was often in town, shopping and picking up her mail at the post office, which meant waiting in a lengthy outdoor queue, while all of Tombstone walked by. Harry Jones, who acted as Wyatt's lawyer on at least one occasion, may have introduced her to Wyatt. She was publicly seen on Johnny Behan's arm, and Wyatt would have known that the alluring Josephine was living with his professional rival. But now the enmity between Wyatt and Johnny would become personal. Just because Behan was cheating on Josephine did not mean he was ready for another man to have her, especially Wyatt Earp. As Wyatt's biographer Stuart Lake put it: "In back of all the fighting, the killing and even Wyatt's duty as a peace officer, the impelling force of his destiny was the nature and acquisition and association in the case of Johnny Behan's girl. That relationship is the key to the whole yarn of Tombstone."

All signs point to the likelihood that Josephine and Wyatt began an affair in Tombstone. The location of their trysts is unknown, but Josephine was surely angry enough to have enjoyed cuckolding Behan in the house they once shared. Young Albert Behan would have been an important consideration, but if his presence was inconvenient, two passionate, determined lovers could find other places to meet.

Decades later, Stuart Lake would say privately that "everyone" in town was buzzing about the former Mrs. Behan and faithless Wyatt Earp. However, Josephine and Wyatt were quite discreet. There was no general awareness that they were together, or even that Josephine had broken up with Behan. The local newspapers covered politics rather than scandals, so the tangled love life of Sheriff Johnny Behan and Josephine Marcus remained their own private problem. It would take years before Josephine adopted the name Mrs. Earp—Tombstone already had quite a few of those. Allie Earp was skeptical about whether Josephine had even been in Tombstone, let alone that Josephine and Wyatt were lovers there. Early Tombstone chroniclers such as William Breakenridge and Walter Noble Burns never mentioned Josephine—but then again, with rare exceptions, they mentioned no women at all.

Tombstone had other things on its mind during that endless hot summer, when the temperature often hit 110 degrees in the shade and the wind kicked up an acrid dust that left an unpleasant coating on the tongue. Aside from the weather, the newly formed county was experiencing an unprecedented crime wave that kept the city's lawmen almost too busy to worry about their love lives.

Cochise County swiftly became a hotbed of contradictions and jurisdictional challenges. Social, economic, and political rifts opened up, as deep and unstable as the silver mines that yawned below the streets of Tombstone. The wounds of the Civil War remained divisive, with battlefield memories still fresh in the minds of former soldiers. In the Arizona Territory, many cowboys or their families had fought for the Confederacy, and were displaced by carpetbaggers after the war. On the other side were Republican families like the Earps, where three brothers had fought for the Union. There was no middle ground among Tombstone residents: they were committed to one faction or another. The political landscape aligned with economic interests: the Republicans were townspeople who supported law and order to protect investments, and stood in staunch opposition to mostly Democratic rural ranchers and cowboys, who cherished their individual freedoms. The town's two leading newspapers, the *Epitaph* (Republican) and the *Nugget* (Democratic), advocated ceaselessly for their parties. Partisan battles broke out over decisions large and small.

The town's peacekeepers proved incapable of setting aside their differences. As representatives of federal and local law, the Earps should have cooperated with county sheriff Johnny Behan. Instead, each side eagerly subverted the other and allowed personal and professional jealousies to compromise their judgment. Behan had the job Wyatt wanted; Wyatt had the girl Behan had brought to Tombstone.

As a series of serious crimes unfolded in Tombstone, the stalemate between the two sides slowly heated up into an all-out war over a high-stakes prize of money, power, and Josephine Marcus.

THE ROAD TO the O.K. Corral began with a sequence of thefts and killings that brought the Earp and Behan factions into open opposition. An audacious attack on a Wells Fargo stagecoach going from Benson to Tombstone left the driver and one passenger dead, and a cargo of silver worth $25,000 missing. Stealing the U.S. mail in addition to the silver kicked the crime into the higher category of a federal offense, and brought deputy U.S. marshal Virgil Earp and Pima County deputy sheriff Wyatt Earp into the picture. Suspecting the involvement of the cowboys, the Earps set off on horseback at midnight, with the county forces, led by Johnny Behan, in another posse. The Earps caught one suspect and turned him over—with considerable reluctance—to Sheriff Behan. Their worst suspicions were confirmed when the prisoner casually walked out of jail a few hours later and disappeared into the streets of Tombstone.

Wyatt dearly wanted to make a couple of spectacular arrests before the next election, when he might wrest the sheriff seat from Behan. His campaign platform was simple: only Earp could end the crime spree. There was also money at stake, since Wells Fargo had posted a handsome reward for the Benson killers. Wyatt offered the Wells Fargo bounty as a bribe to Ike Clanton, a friend of the thieves, if he would lure the robbers, dead or alive, into Wyatt's trap. If Johnny's corruption could be exposed at the same time, Wyatt would move ahead on all fronts—with the bonus of clearing the name of his friend Doc Holliday, who had been accused of taking part in the stagecoach robbery by none other than his lover, a very drunk Big Nose Kate.

The town's collective nerves were set even more on edge when an outbreak of Indian hostilities was reported. "Our city is in a big excitement at present, the Indians are on the war path all around us and have killed a great many people," Louisa Earp wrote breathlessly to her sister. "The people here expect to be attacked any day. Passenger trains in Benson were halted, probably to bring in soldiers." But then the threat of Apaches on the warpath was eclipsed by yet another spectacular crime.

In early September a second stagecoach was held up and robbed of its mail and money, this one on its way to Bisbee. Again, federal and county groups competed for the bragging rights of bringing in their prisoners. This time two suspects were arrested: Pete Spence and Frank Stilwell. Both were known associates of Johnny Behan.

As Cochise County sheriff, it was Johnny's job to keep the cowboys in check. Instead, he always seemed to be looking the other way. Tombstone mayor Clum said that Johnny "winked at crime." But less partisan observers like Endicott Peabody took a more nuanced view. Peabody blamed Behan for the growing cowboy influence, but suspected that the Earps were more than a little ambivalent about helping him: "The cowboys are a troublesome element and, there is unhappily, a feud between them and the marshal's party. The misfortune is that the cowboys are countenanced by the sheriff for political reasons and the marshal's party on the other hand is not quite above suspicion."

Acting governor of Arizona John Gosper conducted an independent

investigation. While he too found fault with both sides, the "cow-boy element at times fully predominates," he concluded. With the tacit consent
of Sheriff Behan, cowboys were running rampant, stealing cattle, holding
up stagecoaches, and attacking federal property. The governor listened to
Johnny's complaints about Virgil, and then heard Virgil leveling the same
charges at Behan. His summary: "The very best law-abiding and peace-
loving citizens have no confidence in the willingness of the civil officers
to pursue and bring to justice that element of out-lawry so largely disturbing the sense of security." Neither Johnny nor the Earps had shown
themselves capable of compromise or coordinated action on behalf of their
constituencies.

"Something must be done, and that right early," Gosper predicted, "or
very grave results will follow."

Gosper's dire warning was the catalyst for the formation of a Tombstone Citizens Safety Committee to work around the town's peacekeeping
paralysis. In case of emergency, an alarm would sound, and the one hundred or so committee members would gather at an appointed place. It was a
vigilante solution to a failure of government.

ON THE MORNING of October 26, 1881, extreme fatigue and painful hang-
overs befuddled the brains of an unusual gathering of men who had been
gambling and drinking all night. The unlikely group around the table
included Virgil Earp, Johnny Behan, and his friends and cowboy sympathizers Ike Clanton and Tom McLaury. The game was anything but
friendly, as the participants had a long list of reasons why they hated each
other. Johnny was the least experienced gambler among them, an easy
mark for the professionals. But he had come to the table determined to win,
knowing that there was a lot more on the table than cash.

It was almost daylight when the game broke up. Long after the rest of the
poker players staggered off to bed, Ike Clanton continued to roam the streets

of Tombstone, waving his gun and complaining loudly about the Earps and Doc Holliday. As chief of police, Virgil showed up and answered Ike by hitting him over the head and arresting him. Ike was not supposed to carry a gun, since Tombstone's "Ordinance Nine" had been enacted to prevent armed confrontations in town. Although this regulation was, at best, applied inconsistently, Virgil acted swiftly. Ike was dragged into court and walked out with a still more throbbing head and a $25 fine for carrying a weapon.

Another one of the poker players that night, Ike's friend Tom McLaury, took up the war of words and was the next one to get whacked over the head, this time by Wyatt Earp.

Each side gathered reinforcements. Ike kept up his steady stream of schoolyard taunts. He and Tom were joined by their brothers, Billy Clanton and Frank McLaury. Virgil huddled with Wyatt and Morgan. Doc Holliday, who had heard enough of Ike's loud accusations, joined them. The chorus of anger and aggression grew louder, the volatile brew of jealousy and political ambition and greed and historical resentment more toxic. Exhaustion clouded the capacity of both sides for mature judgment. October 26 was a day of sneering, taunting, verbal and physical abuse, dares and double dares, and finally, deadly provocation. The Earps closed ranks and responded with an unprecedented display of police firepower.

Johnny Behan's job was to keep the peace, and he failed. He slept late that morning, and then he indulged in a leisurely shave, thus missing most of the day's provocations. Virgil Earp's job was to keep the peace, and he also failed. There were hours in which either one of them could have defused the hostilities. Behan was indecisive and flustered in the face of escalating tensions. As Walter Noble Burns put it, Behan was "a man of words rather than action.... [He] talked like a big man and acted like a little one." Instead of recognizing that personal resentments were compromising his judgment, Virgil Earp gathered his deputies—including the most unlikely peacekeeper, Doc Holliday—and confronted his enemies in the middle of Tombstone's afternoon bustle.

With the rest of Tombstone going about its business, buying groceries, cashing checks, and sharing the news of the day, Virgil, Wyatt, and Morgan Earp and Doc Holliday confronted Ike and Billy Clanton, Frank and Tom McLaury, and Billy Claiborne in an open lot near the O.K. Corral. The next few minutes were a blur of shouts and shots and flying bodies—a complex tour de force of sheer drama that has fueled more than a century of controversy.

Who shot first?

Were the cowboys armed?

Did the Earps act in self-defense, or with homicidal intent?

These and a hundred other questions have been asked and answered, and then asked again. Instead of answers, there is only the terrible beauty of the three handsome Earp brothers and crazy Doc Holliday stalking the dusty streets of Tombstone in their long black coats, approaching their enemies with guns loaded and a grim determination to finish a bad day's work.

"YOU ALREADY HEARD about these events by telegraphic, so I won't repeat but will add extra details," Clara Spalding Brown informed her readers in San Diego a few days later, and then went on to describe the scene:

> *Following the Indian scare, stage robbery, and important mining suits, we have been shocked by the most tragic and bloody occurrence which has transpired in the history of the camp. . . . The inmates of every house in town were greatly startled by the sudden report of firearms, about 3 p.m. There were so many shots that it sounded like fire-crackers, followed by a shrill whistle coming from the mines. "The cowboys!" cried some, thinking that a part of those desperadoes were "taking the town." "The Indians!" cried a few of the most excitable. Then the whistle sounded again and well-armed citizens appeared from all quarters, prepared for any emergency.*

Josephine heard the thunder of the gunshots and the piercing shriek of the whistle summoning the Citizens Safety Committee. Without stopping to put on her bonnet, she ran toward the scene as fast as she could, expecting to find Wyatt there. With the uproar in the streets, she would have been able to slip unnoticed into the first ring of men and women gathered there. Once she saw that Wyatt was alive, she melted back into the crowd and retreated.

Wyatt was the only man left standing. Bodies were removed in a chaotic scene of gore and confusion. A bullet had grazed Doc Holliday. Morgan and Virgil were injured more seriously. On the cowboy side, no medical attention was necessary, other than morphine to ease the last moments of Billy Clanton. Frank and Tom McLaury were already dead. Ike Clanton had escaped harm when Wyatt, who could have shot Ike in an instant, shoved him to safety. As many would note afterward, Wyatt's remarkable restraint was the strongest evidence that he did not set out to murder unarmed men.

Johnny Behan rushed to the scene and took charge. The bodies of the McLaurys and Billy Clanton were carried to a nearby funeral home, where autopsies were performed.

Virgil and Morgan were put to bed together, attended by the "misses Earp": Allie, Louisa, and Mattie. Like everyone else in Tombstone, the women had heard the alarm but knew little of its significance. "We never realized what things were comin' to," Allie said. "The men didn't talk much about it at home for fear of scarin' us I guess." They had no reason to suspect that Wyatt, although unharmed physically, was already dead to Mattie.

The cowboys' funeral was the most elaborate ever seen in Tombstone. Led by the Tombstone Brass Band, hearses and wagons followed with the rest of the Clanton family, and a long trail of carriages and horsemen brought up the rear. "Almost like a fourth of July celebration," Allie Earp sniffed in disgust. Hundreds of cowboy sympathizers walked behind the procession, with another thousand people lining the streets. Clara Brown noted the careful formality of the occasion, the bodies "handsomely laid out" in a manner that she considered incongruous to the circumstances. "A stranger viewing the funeral cortege . . . would have thought that some person esteemed by the entire camp was being conveyed to his final resting place," she observed. "Such a public manifestation of sympathy . . . seemed reprehensible when it is remembered that the deceased were nothing more or less than thieves."

Clara Brown voiced a point of view initially held by many in Tombstone's business establishment who stood firmly behind the Earps and pointed a harsh finger at Behan, accusing him of acting in collusion with the cowboys, even to the point of accepting bribes. But she was troubled when reports surfaced that at least two of the cowboys may not have been armed. At the very least, she acknowledged two sides to the controversy. "Opinion is pretty fairly divided as to the justification of the killing," she considered. "You may meet one man who will support the Earps, and declare that no other course was possible to save their own lives, and the next man is just as likely to assert that there was no occasion whatever for bloodshed."

As Clara predicted, Tombstone became a "warm place" for the Earps. After a contentious year of conflict and provocation, frontier justice moved swiftly through a thicket of hearings, writs of habeas corpus, arrest warrants, indictments, affidavits, remanded testimony, and jurisdictional disputes. A formal inquest was convened two days after the gunfight. Ten jurors considered the evidence and failed to find against the Earps and Holliday. In disgust, Ike Clanton filed a first-degree murder complaint. Morgan and Virgil, still bedridden, were suspended from duty. Wyatt and Doc were taken into custody. Bail was fixed at $10,000, and was raised by Wyatt's friends. Doc would probably have languished in jail had Wyatt not urged his friends to help him. Big Nose Kate Elder, contrite over her false accusations, gave Doc most of the $100 she possessed and left town.

Attorney Tom Fitch, an experienced courtroom lawyer, helped to raise bail for Wyatt and undertook his defense. Fitch's strategy depended on

winning the hearing, and thus avoiding a trial in which Wyatt might con-
front a jury full of cowboys. Frank and Tom McLaury's brother, successful
lawyer Will McLaury, rushed from Texas to aid in the prosecution of his
brothers' killers. "I think we can hang them," he assured his sister. At his
insistence, bail was revoked for Wyatt and Doc. But the prosecution never
recovered from the disastrous testimony of Ike Clanton. He was an unsym-
pathetic witness, rambling and inconsistent, and spoke in a high falsetto
voice, "rather like a rusty hinge," as one observer noted. Fitch outmaneu-
vered the prosecution by demonstrating that Ike had entered into a secret
agreement with Wyatt to hand over the Benson stage robbers in exchange
for the reward.

Fitch pushed Johnny to repeat his grievances against Wyatt and caught
him in several lies. Behan testified that he had originally intended to appoint
Wyatt as deputy sheriff, but he evaded any explanation of his change of
heart, saying only "something afterwards transpired that I did not take
him into the office." He also denied that he had any particular relationship
with Ike Clanton; it would be decades before it was known that Behan had
secretly underwritten Clanton's court costs.

Wyatt's testimony was far more effective. He used his physical presence to his advantage and read in a sonorous voice from a well-prepared
script. Without apology, he disclosed his ambitions to become sheriff: "I
thought it would be a great help to me with the people and businessmen if
I could capture the men who killed [the stagecoach driver]. . . . I told them
I wanted the glory of capturing [the robbers] and if I could do it would
help me make the race for Sheriff at the next election." He also presented
impressive character references from his former constituencies in Wichita
and Dodge City.

There were two sides at the hearing, and Josephine was significant to
both of them, but her name was never spoken, her existence never revealed.
If Fitch or Will McLaury knew about her role in fueling the hatred between
Wyatt and Behan, they remained silent. Introducing her story would bring

no obvious advantage to either side. It would certainly not have helped Behan to admit his infidelity and jealousy, which would have then been considered as additional motives in his war against the Earps.

After a month of testimony, Spicer issued a strong verdict acquitting the Earps. It did not matter who shot first, he reasoned, only that the Earps had acted in "discharge of an official duty." He invited the grand jury to consider new indictments if they disagreed with the decision, but in the meantime, "there being no sufficient cause to believe the within named Wyatt S. Earp and John H. Holliday guilty of the offense mentioned within, I order them to be released."

THE EARPS SHOULD have left Tombstone. While they had been vindicated, the outcry from the other side made it clear that Spicer's decision would hardly stand as the last chapter in the Earp-cowboy war. Virgil and Morgan were now strong enough to travel. Their family in California urged them to come back. Their professional future as lawmen was cloudy: acting governor Gosper had visited Tombstone during the Spicer hearing and believed that Johnny Behan and the Earps should be permanently removed from office.

But Wyatt resisted the pressure to leave. The brothers had substantial business interests, including a part ownership in the Oriental Saloon and mining claims such as the "Mattie Blaylock," which Wyatt had filed soon after they arrived in Tombstone. He held on stubbornly to the hope that he might run again as sheriff and defeat Behan. Josephine was still in town.

The "divided state of society in Tombstone" that Clara Brown observed at the cowboys' funeral intensified, stoked daily by the town's two newspapers. The cowboys were rumored to be holding secret midnight vigils, swearing blood oaths against the Earps, who represented everything in government that they most hated. Death threats were made against Spicer and Mayor Clum. Far from having any gratitude toward Earp for saving his life during the gunfight, Ike sought another day in court with new legal representation; Will McLaury left town in disgust.

Tombstone business interests cared less about justice than they did about resolving controversy and bad press that interfered with the flow of capital. The town's natural riches could make the county so great, Clara wrote, but for "an element of lawlessness, an insecurity of life and property, an open disregard of the proper authorities, which has greatly retarded the advancement of the place." In her view, Tombstone was as safe as San Diego or Boston, but the town was getting a terrible reputation. Around the country, leading national newspapers were writing about the gunfight. Even in Australia, there was talk of a "cowboy war" raging in Tombstone between a family of professional gamblers named Earp and a "tribe" known as the Clanton gang. "Goodness knows when it will end," reported the Australian *Town and Country Journal.*

The Earps should have left.

"Ever since our trouble we have stayed at the Cosmopolitan Hotel and it is very disagreeable to be so unsettled," Louisa wrote to her sister. Wyatt had heard rumblings of a cowboy hit list and moved the family out of their houses and into the Cosmopolitan, while the cowboys took up residence at the Grand Hotel across the street. The "trouble" had also disrupted the Earps' businesses, especially now that Ike had begun a new lawsuit. To raise money to cover their legal fees, Wyatt sold his share of the Oriental Saloon and mortgaged his house in a legal document that bore his name as well as that of "Mattie Earp, wife." But the clock was ticking down the days of their marriage.

The "Cowboy and Earp feud," as Endicott Peabody and Clara Brown called it, exasperated the entire town. Ike was defeated a second time in court, but he renewed his public demand for revenge. He'd lost his brother and two friends, his public reputation was shredded, and he had been twice thwarted by the justice system. George Parsons recorded the judge's decision with a diary entry that "a bad time is expected again in town at any time. Earps on one side of the street with their friends and Ike Clanton [with his supporters] on the other side—watching each other. Blood will surely come."

On December 28, Tombstone exploded again. Under the headline "Shooting a US Marshal," the *New York Times* reported that "Deputy United States Marshal [Virgil] Earp was fired upon while crossing Fifth Street last night by three men, armed with shotguns, who were concealed in an unfinished building and who escaped in the darkness." The hail of bullets shattered Virgil's elbow. According to George Parsons, Virgil was carried away on a stretcher while reassuring a terrified Allie: "Never mind, I've got one arm left to hug you with." Virgil begged Wyatt to stop the doctor from amputating the shattered limb.

Josephine was still living quietly somewhere in Tombstone. With Wyatt's every move being watched carefully by the Clanton faction, the lovers would have had little opportunity for a liaison or to make plans for the future. Even exchanging greetings would have drawn Josephine into more danger. She could do nothing but wait.

DESPITE THE EXPLICIT warnings of the Arizona governor to President Chester Arthur, outlining the persistent "problems of lawlessness" in southeastern Arizona and emphasizing that neither the Earps nor Johnny Behan deserved any law enforcement authority, Wyatt received an appointment as a deputy U.S. marshal to hunt down Virgil's assailants. He was wearing a badge again.

Wyatt kept the family indoors and guarded. Virgil had not left his bed at the Cosmopolitan Hotel. Morgan, however, was angling for a night out. Wyatt relented and agreed to accompany his brother to the opening night of a comedy at Schieffelin Hall. Surely a night of laughter would do them both some good, reasoned Wyatt. Surprisingly, this cautious man of few words loved the theater. Allie delighted in telling the story of how Wyatt once became so engaged in a performance that he stood up and threatened to draw his gun in defense of a young actor whose fictional character was falsely accused. "That was Wyatt for you," Allie recalled. "He wasn't the one to stand by and see wrong done to an innocent boy anytime." But on March 8, 1882, he should have obeyed his instincts. Instead of staying safely at the hotel, Wyatt and Morgan went to see *Stolen Kisses* and then finished up the evening with a game of billiards at Hatch's, a nearby saloon.

The game had just begun when two rifle shots rang out: one missed Wyatt, but the other tore into Morgan's right side. He lived for one more hour. Allie Earp had ventured out of the hotel to buy taffy for Virgil, and from a few blocks away she saw the commotion at the pool hall. Tiny Allie crawled between the legs of the men clustered around Morgan to see who had been shot. Wyatt saw her and said, "Allie, you and Doc [Goodfellow] fix up Virge so he can get out. Then I can go get those fellows."

Now Wyatt had also lost a brother.

Morgan's inquest drew a most unexpected informant. Josephine's friend Maria testified against her husband, Pete Spence. It had been just over a year since she and Josephine shared gumdrops on the Benson stagecoach. The once-hopeful brides had remained friends in Tombstone, linked by their stagecoach trip and by mutual friends such as the local milliner. Perhaps they had cried on each other's shoulders for their bad choices of husbands, though Maria's match was arguably even worse. Pete was beating Maria and her mother, who was living with them.

Maria spoke out against her husband, telling the inquest jury that they had quarreled after she saw him identify Morgan to another member of his gang as the intended target of an assassination plot. Pete struck her and threatened to shoot her if she told anybody about what she had seen. After she testified, Maria privately warned the Earps that even now, with Morgan dead, the violence was not over. In her memoir, Josephine dramatized Maria's bravery at the most critical moment of Wyatt's life: "A dark-eyed young woman—the wife of the outlaw, Pete Spence—slips into Addie Bourland's [sic] millinery store—and saves the Earp family from annihilation."

Morgan's body was taken by train to his parents in California, accompanied by his brother James. A few days later Virgil and Allie followed, guarded by Wyatt and a small army of his friends. They were expecting trouble. Allie was wearing Virgil's six-shooter under her loose coat, sitting at his right hand so he could draw the gun at the first sign of trouble. The train pulled into Tucson. While it idled at the station, Wyatt glimpsed the ambush he expected: Frank Stilwell and Ike Clanton, lurking on the tracks. When Wyatt followed the men, Ike ran away, as he had in Tombstone, but Wyatt caught up with Frank Stilwell. Without bothering to be discreet before or after, Wyatt unloaded so many bullets into Stilwell's body that one observer called him "the most shot up man I ever saw."

For the rest of her life, Allie would remember those tense hours at the Tucson train station and Wyatt running alongside the train, holding up a finger and shouting, "It's all right, Virge. We got one! One for Morg!"

Knowing that Wyatt was hell-bent on revenge, Allie remembered that they "kept worryin' about Wyatt" as the train left for San Bernardino without him. He was the only Earp left in Arizona. It would be years before they saw each other again.


MATTIE AND JOSEPHINE were the last of Wyatt's entanglements still in Tombstone. There was no room for romance in Wyatt's life, not now. He put Mattie and James's wife Bessie on a train to join his family in San Bernardino. With a compassionate farewell, the *Tombstone Epitaph* noted the departure of the two Mrs. Earps: "They have the sympathy of all who knows them." Josephine's departure went unnoticed.

Wyatt Earp may have been devoid of physical fear, but he sent Mattie off to his parents without an honest word of good-bye.

FRANK STILWELL'S KILLING marked the beginning of what became known as Wyatt Earp's Vendetta Ride.

It was Wyatt's actions after Tombstone, at least as much as the Gunfight at the O.K. Corral, that sealed his reputation as an implacable avenger. Now he would ride for justice, not for law. He had seen enough of courtroom procedures to conclude that nothing made sense; after all, his attempt to represent the interests of businesses against the unruly and individualistic world of the cowboys had led to this grievous wound to his family. The courts had failed him. Henceforth, justice would be swift, personal, and final.

Josephine would only learn the details much later, and not from Wyatt. His friends told her how Wyatt and Johnny Behan led different posses around the Arizona Territory for months, sometimes within sight of each other, Behan waving a warrant for Wyatt's arrest and making a great show of following him, but only halfheartedly trying to capture his dangerous prey, while Wyatt ignored Behan and continued to hunt his brothers' attackers, one by one, with ruthless precision. Again and again, Wyatt emerged unscathed from savage gunfights.

The Vendetta Ride was followed closely, especially in Tombstone, where public opinion was divided as to the righteousness of Wyatt's quest. "The sheriff was at the head of a gang of cowboys, hunting the Earps. A nice community," was Clara Brown's sarcastic summation. "The posses have marched and counter-marched until people have become accustomed to seeing armed horsemen upon the streets." She considered it unfair that newspapers like her own *San Diego Union* carried lurid accounts of the Frank Stilwell murder, with no mention of the "even more villainous" assassination of Morgan Earp that preceded it. That Wyatt had turned his back on the law was inexcusable, but she urged her readers to consider that "it was not the Earps who first disturbed this quiet, and that their criminal actions since have been from the determination to avenge the murder of a dearly beloved brother." George Parsons had no sympathy for Stilwell and his associates, and congratulated Wyatt on "a quick vengeance, and a bad character sent to Hell."

As Wyatt's body count increased, his advocates back in Tombstone began to waver in their support. The town's future was at stake: Would it be remembered for its success as a mining town, or as the seat of terror and anarchy? Even Earp partisans had grown weary of the tension created by the vendetta. "Fine reputation we're getting abroad," Parsons worried. He was ever the most bellicose of the Earp supporters, but he had investments to protect. "A regular epidemic of murder is upon us," declared Endicott Peabody, fearing that the town's violence would compromise his attempts to expand his congregation and build a new church.

By April, the killing was over. The groups on horseback broke up, and the exhausted participants staggered back to Arizona, Colorado, and New Mexico.

Wyatt's next steps were uncertain; while the family still had business interests in Tombstone, including the Mattie Blaylock mine, not even he could doubt that the Earps were finished in Tombstone.

Johnny Behan was done, too, though it would take more time before he left Tombstone. As with Wyatt Earp, opinion would remain sharply divided about Behan's character and motivation. He gained a coveted place in history as the first sheriff of Cochise. His deputy Billy Breakenridge described him as a "brave and fearless lawman, who could see something good in even the worst of men." But Earp advocates such as Sol Israel mocked Johnny as a man who was afraid of his own shadow. "The Tombstone pace was a bit too fast for him," scoffed mayor John Clum.

Behan never wore a badge again, but he did have a respectable, even colorful career after Tombstone. As a civil servant he was posted to Tampa, then Havana, and lived in China during the period of the Boxer Rebellion.

He had senior political appointments as a customs inspector for the port of Buffalo and warden of Yuma Territorial Prison. He never won another election or married again; his advancing case of syphilis compromised his health, and eventually killed him.

It was Wyatt, more than Johnny, who remained a controversial figure, forever defined by Tombstone and the Gunfight at the O.K. Corral. From that time on, the name Earp would set off a chain reaction that would make some cheer, some cringe, and all look around to make sure that there was no gun pointed in their general direction.

Tombstone marked the end of the Earp brothers' dream of shedding their law enforcement jobs and becoming a clan of capitalists. The brothers were not given to recrimination against one of their own, nor did they dwell on the past. There was enough blame to go around, but it was Wyatt who had kept the brothers in Tombstone too long, in the mistaken belief that he could win an election against Sheriff Behan. He wanted to leave town with Josephine, and not with Mattie; instead, he left with neither. The family had suffered much, and gained little, in Arizona.

AFTER THE DRAMA of the Vendetta Ride, a much ballyhooed visit from the commanding general of the U.S. Army, William Tecumseh Sherman, provided a welcome distraction for the people of Tombstone, who must have wondered whether to be gratified or insulted by his surprise to find "such a number of fine looking intelligent citizens in this place so badly thought of outside."

About six weeks after Sherman's visit, another fire struck the beleaguered city, and this time the destruction was still more devastating. The long-suffering citizens of Tombstone had rebuilt before, and they were sure that they would rebuild again, but now the town's foundation was shaken to the core. The recent history of lawlessness, constant fires, and threats of Indian attack were all factors in Tombstone's demise, but what finally doomed the town was that the silver was running out, and the mines were filling with water. Labor unrest was growing, and wages were falling. The railroad came tantalizingly close . . . and then terminated, less than fifteen miles away.

Those who had come to Tombstone with vague notions of adventure and wealth began to leave, most of them no richer than before. Gamblers, prostitutes, and saloonkeepers moved on to ply their trades in the next boomtown. By the fall of 1882, even some of the town's leading citizens were moving on, including the memorable trio of Peabody, Brown, and Parsons.

Endicott Peabody never had intended to stay very long, and he grew bolder in his dislike of the place. "O unfortunate! To contemplate a sojourn in such a country," he warned a prospective settler. The inhabitants are "rotten," he said, and life there is "crude," dismissing the whole lot of Tombstoners in his sweeping indictment. Yet he never wavered in his good opinion of Wyatt Earp. "YOUNG MAN," elderly Peabody raised his voice in rebuke of a whippersnapper who inquired if the distinguished educator and clergyman really approved of "such men" as Wyatt Earp. "I don't think you realize the kind of person we needed as law officers 60 years ago," Peabody lectured. "The Earps were very good law officers."

Clara Brown was the next to depart. She and her husband had lost their nest egg after the destruction of their bank in the June 1881 fire. "All feel that this is a place to STAY in for a while; not a desirable spot for a permanent home," Clara predicted early in her residence. Now that it was time for her to leave, she vowed to remain a booster, declaring that Tombstone still had a bright future as a good place to make money, despite being "so gloomily named a town."

The last to leave was George Parsons, who stayed for a few more years before moving to California. For the rest of his life, he would defend the Earps as "a benefit and a protection to the community." He would continue to cross paths with Wyatt in far-flung places, and would remain his loyal admirer, though he remained resolutely silent on the subject of Josephine Marcus.

Tombstone deflated as quickly as it had expanded. When the biggest mines closed down in 1886, half the population fled in just six weeks.

For Josephine and Wyatt, Tombstone would remain forever the wellspring of their romance, despite the violent contradictions of that complicated time when she belonged to Johnny Behan and fell in love with his rival, and when Wyatt's family peace was shattered forever.

For now, she was Josephine Sarah Marcus again, back in her parents' home in San Francisco.

Mattie Blaylock Earp was in San Bernardino, California, with the other Earp wives.

Both of them were waiting for Wyatt.

## 2 | THE FOURTH MRS. EARP

$\mathcal{J}$OSEPHINE'S SECOND return home was as ignominious as her first. From her parents' point of view, she accomplished nothing in Arizona other than embarrassing love affairs. She had been in contact with her parents during her brief career as Mrs. Behan, but it must have taken some serious talking to explain that she was now in love with Behan's archenemy, the subject of so many lurid stories in their local newspaper.

The contrast between her home in San Francisco and the lethal but exciting world of Tombstone was even more jarring to Josephine than before. Yet her family hoped that she would settle down, and perhaps be more like her siblings, who had done a much better job of meeting their parents' expectations. Josephine's older half sister Rebecca had married Aaron Wiener and was raising a Jewish family. Nathan was still in school, though he was not much of a student and seemed to have little ambition. Younger sister Hattie was a charming young girl of seventeen, with large blue eyes and light hair. An up-and-coming businessman named Emil Lehnhardt, who was also from an immigrant family, was courting her, and so far he had expressed no strange plans to conquer the frontier West.

It was hard for Josephine to predict what her future might be with Wyatt, but it wasn't likely to fall into a pattern that anyone in the family would consider normal.

No one would blame Josephine's mother and father for wanting to lock her in her bedroom. Within a few weeks of having seen one brother assassinated in front of his eyes, and another crippled for life, Wyatt had killed at least three men. During his reign of terror, he lived for months in a shadowy jurisdiction where he was both outlaw and lawman, his posse circling around Behan's like hungry wolves. By the end of the Vendetta Ride, Wyatt and his friends were confronting the more mundane work of figuring out their futures. They left their horses in Arizona and crossed into New Mexico. There was money to be divided—the remainder of the funding Wyatt had received from the U.S. marshal and Wells Fargo—and debts to be settled. Wyatt stayed with his friend Henry Jaffa, a prominent Jewish merchant from Colorado and New Mexico, who gave Wyatt a coat to replace the one he was wearing, now full of bullet holes and covered in the filth of the trail.

Wyatt and his men had been on horseback for a long time, and tempers were short. Doc Holliday caused the most commotion; he had an irrepressible tendency to rage, especially when he was drunk, which was most of the time. One night, he went too far. Wyatt was becoming "a damn Jew boy," he complained. Doc also hinted that he had given money to "Behan's woman," which would later fuel speculation that Doc had been intimate with Josephine, though he might also have been referring to Behan's expensive prostitute, Sadie Mansfield.

Doc's ill-timed insult was a distraction that soon blew over. But it signaled bigger problems that would soon divide the friends forever, with Josephine again in the middle.

Behan was pushing for Wyatt's extradition back to Arizona to stand trial for the murder of Frank Stilwell. Colorado governor Frederick Pitkin was pressured from all sides, but was ultimately swayed by George W. Crummy, an influential saloonkeeper who allegedly ran gambling houses in partnership with Pitkin and may have been an associate of Bat Masterson. Together, they managed to lift the Stilwell indictment permanently. If Bat could have helped Wyatt without assisting Doc, he probably would have, as Bat had little respect and no affection for Doc. But the cases were inextricably linked, and so both Wyatt and Doc benefited from Bat's timely intervention.

With the legal situation settled, Wyatt's Tombstone chapter was closed. Turning his back on Mattie, Wyatt headed for San Francisco. Virgil was there, consulting with doctors about his shattered elbow and meeting with Wells Fargo, who presented him with a large gold five-pointed star for his efforts to keep the peace in Tombstone.

Josephine was also there, waiting for Wyatt. Whatever commitments he offered were enough for her. Without any wedding ceremony, Josephine was ready to leave San Francisco with him.

Perhaps her parents had grown accustomed to Josephine's bewildering changes of suitors. Wyatt had none of Johnny Behan's unctuousness, and unlike Aaron Wiener or Emil Lehnhardt, he had little to offer in the way of a secure future. At least Johnny Behan had presented himself as a lawman with some business interests. The glibness that greased his path through politics would have come in handy when wooing prospective in-laws.

Between Wyatt's compelling physical presence and Josephine's obvious determination to attach herself to him, Hyman and Sophia Marcus probably made little attempt to stop Josephine from leaving home a third time. Of all their children, Josephine would always be the most unconventional.

JOSEPHINE BEGAN THE next phase of her life in Utah, excited about leaving San Francisco with Wyatt and staying in elegant hotels. In her memoir, she invested some of her first encounters with unusual significance. In Salt

Lake City, she listened thoughtfully to an impromptu tale of woe from a Mormon hotel maid, who poured out her heart about her childless marriage. Her husband cast her aside for a second, and then a third wife, until at last one of her rivals delivered a child. From Utah, Josephine traveled to Colorado, where she stayed at the luxurious Tabor Hotel, owned by a wealthy industrialist and his beautiful young wife, the improbably named Baby Doe. Although she could not have known it at the time, the glamorous Tabors would come to a sad end, and Baby Doe would die in despair and debt. Josephine would have much in common with both women: the barren maid and the impoverished widow.

But for now, Josephine cared only for her immediate happiness. After her furtive days in Tombstone, she was exhilarated to be openly acknowledged as Wyatt's woman by his closest associates, especially Bat Masterson and his wife Emma. They made a congenial foursome. Emma had also taken a turn on the stage, and she and Josephine became good friends. Josephine was immediately taken with Bat Masterson's "Irish blue eyes and long curling eyelashes" and listened eagerly to his explanation of why Wyatt felt compelled to "clean up" the frontier by comparing him to a "fastidious housewife" who "hates dirt and untidiness in her home." For his part, Bat openly admired Josephine and compared her to the celebrated actress Jefferys Lewis, who happened to be performing nearby, and joined them one night for dinner. Josephine professed not to see the resemblance, other than the basics of another "dark-eyed young woman with smooth skin and rosy cheeks," but she glowed under his flattery.

The Earps traveled alone to explore the silver fields of Idaho, via Ouray, Colorado, and then across the Rockies. Anticipating the rigors of crossing the 11,000-foot Red Mountain Pass, still covered in snow despite the late spring, Wyatt suggested that Josephine exchange her skirts for overalls. She was horrified at first—what would the hotel guests and employees think!— but finally agreed. Over her trousers Josephine added layers of sweaters and a heavy dark coat, scarf, and gloves, all in different blue tones, and then climbed aboard a frisky mule for the long walk over the rocky ridges. A parting gift from Mrs. Masterson, Dickie the bright yellow canary made the crossing too, tenderly fed from Josephine's hand and kept warm in a carefully wrapped cage. At the summit they surprised three reporters, two from Chicago and one from Denver. Josephine's initial embarrassment turned to pleasure when they waxed poetic about the apparition:

> *"We couldn't believe our eyes," said one. "Here we've been trudging through this stood-on-end wilderness of snow all day without seeing a soul, when suddenly out of the skies a brown-eyed goddess appears—"*
> *"Riding on a donkey—" added the second.*
> *"With a blond Apollo to push it along," the man from Denver finished.*

The reporters were still more tantalized to discover that the "blond Apollo" was actually Wyatt Earp, whom they believed to be a fictional character from a Western thriller. The reporters kept their promise to send their article, a "glowing account" that recorded the chance encounter as a highlight of her early life with Wyatt: overalls, canary, and all. It was not the last time that Wyatt Earp would surprise someone expecting a legend, not a man.

The couple went separate ways in the summer of 1883 to attend to matters of friendship and family. Wyatt returned to Dodge City, Kansas, at the request of his friend Luke Short, who was embroiled in a legal tussle over his right to operate saloons and whorehouses. As part-owner of the Long Branch saloon, Luke felt that he had been unfairly targeted by newly imposed "moral" ordinances, and he called upon his friends, the most famous gunfighters in the West, to show their support by returning to Dodge City. Not a shot was fired. Instead, what would become known as the Dodge City War Peace Commission resulted only in a famous photograph of Wyatt with Bat Masterson, Luke Short, and others. The reunion was a pleasant one for Wyatt, as was the new lesson that he could win battles by using his reputation and his famous stare, and just the glint of a gun.

Josephine went home to San Francisco for the wedding of her sister Hattie to Emil Lehnhardt, happy to exchange her Rocky Mountain overalls for beautiful dresses that showed off her fine figure and the healthy glow of these joyous first months with Wyatt. It was July 5, 1883, barely three months since she had left home, but the pace of her life—Salt Lake City, Denver, crossing the Rockies—must have been startling in comparison to the relatively staid life of her sisters in San Francisco. As the new wife of a pillar of the Oakland community and a director of the Unitarian Church, Hattie was on a solid path to respectability, stability, and prosperity. Her husband had come to California from New York around the same time as the Marcus family, and was building up a candy manufacturing business, which was already prosperous enough for him to present Hattie with her wedding present: a beautiful large house on Telegraph Avenue and Twenty-Seventh Street in Oakland.

Back on the road with Wyatt, Josephine began using the name "Mrs. Earp" as early as December 1883, when she registered at the Washington Hotel in Galveston, Texas. She had discovered a newfound ability to move gracefully between the worlds of her Jewish parents, affluent sister, and frontier husband. However, no matter where she went, so did the shadow of Tombstone. In Fort Worth, Will McLaury glared at Wyatt when they met accidentally: a man of words rather than bullets, Will would not take his desire for revenge beyond legal limits, but his fury was still fresh. Still worse was a close encounter with the previous Mrs. Wyatt Earp. Mattie was reportedly visiting Big Nose Kate in Globe, Arizona, when the new Mrs. Earp passed through.

Mattie Earp had finally figured out that Wyatt was not coming back to her. After an unbearably awkward period of living with his parents, she left Colton. Not much is known about her whereabouts for the next few years. However, she apparently had the bad luck to visit Globe just when

Josephine and Wyatt showed up. Josephine would have found it gratifying when one of the local papers said that "besides being very handsome, [Josephine] is certainly a lady anywhere." For Mattie, the sight of Josephine, young and beautiful, sashaying around in her finery may have been the final blow that triggered her tragic last chapter.

THERE WAS ALWAYS a new boomtown. Some secret signal seemed to emanate from them, drawing prospectors and those that fed off them.

The new mining sensation of the day was Coeur d'Alene, Idaho. The Earp brothers—minus Morgan and Virgil—reassembled there in nearby Eagle City. As they had before, Wyatt and James dabbled in law enforcement, saloon keeping, real estate, and mining. Wyatt was elected the deputy sheriff of Kootenai County and handled legal threats from claim jumping to at least one murder case. He also took on a new role as innkeeper of the popular White Elephant. "Wyatt Earp escorted us to the rear rooms of his large establishment," recalled one guest. "He hailed from Arizona, where several brothers [of the Earps] had been partners, and were cattle kings." Tombstone was hundreds of miles away, but half-truths about Wyatt Earp were following him, hovering close to the surface.

The Idaho boom was short-lived, however, and the Earps were soon on the move again. Wyatt was working for Wells Fargo for the first time since Tombstone. His work took them all around Texas, including El Paso, Austin, and San Antonio. At Laredo, they crossed into Mexico, where Wyatt became so absorbed by a street card game that he failed to notice a pickpocket, who made off with the gold watch that had been a gift from Senator George Hearst. "They've touched me," he exclaimed, chagrined by his own lack of street savvy.

Josephine's acting days were behind her, but she continued to try on various roles. As they traveled through the farmlands of Texas, she romanticized the simple life of the farmers whose fields and modest cabins she observed through the train windows. Could she be a farmer's wife? That required a leap of imagination for the lively consort of an itinerant lawman and gambler. But she also dreamed about being a respectable lady of means, like her sister Hattie.

In the spring of 1885, Josephine and Wyatt were staying in Denver's Windsor Hotel, visiting from Aspen, where Wyatt had invested in a saloon. Tombstone came walking toward them, in the form of Doc Holliday. Shockingly thin, unsteady on his legs, coughing continually, Doc found it difficult to talk, but he had learned that Wyatt was there, and was determined to see him again. Doc would soon be moving to a sanatorium in Glenwood, Colorado, where he would spend his final days, still drinking a bottle for breakfast and dispensing sardonic commentary from his bed.

Doc clung to Wyatt, as if to derive strength from hanging on his arm. Touched by Doc's condition and its immediate effect on Wyatt, Josephine put aside any lingering resentments. This was Doc, whose irrepressible bad-boy behavior had often riled people up against the Earps, but it was also Doc who once saved Wyatt's life in Dodge City.

Wyatt's loyalty to Doc would remain nearly inviolate. Asked decades later in a legal deposition "whether Doc was somewhat of a notorious character in those days?" Wyatt answered memorably: "Well, no. I couldn't say that he was notorious outside of this other faction trying to make him notorious."

It was too late for any of them to pretend that Doc would recover; his tuberculosis had advanced to its final stage.

"Isn't it strange," Wyatt reflected. "If it were not for you, I wouldn't be alive today, yet you must go first."

Doc died on November 8, 1887. He was thirty-five years old.

UNLIKE TOMBSTONE OR Coeur d'Alene, it would be land, not silver or gold, that defined the next boomtown: San Diego. The combination of great climate, excellent harbor, and train service created "the best spot for building a city I ever saw," in the words of Alonzo E. Horton, the founder of "new" San Diego. Backing up his own predictions, Horton bought nearly a thousand acres, and watched property values explode as the population grew from 5,000 to about 35,000 from 1885 to 1887. Railroad rate wars brought prices down to irresistible lows; a round trip from Chicago plummeted from $150 to $1.

It was Virgil who convened the Earp clan in San Diego. He and Allie had visited the city, and as he had in Tombstone, he saw the possibilities immediately and sent for his brother.

For three years Josephine and Wyatt had been nomads, sometimes staying in luxurious hotels but more often in crude boardinghouses in backwater boomtowns that sputtered out after less than a year. San Diego offered the prospect of glamour and sophistication, perhaps a place worthy of a permanent residence. For once, their Tombstone connections elevated their status: San Diego native William J. Hunsaker had once chased the dream of silver in Tombstone but was now a prominent lawyer back in his hometown. Wyatt also connected again with his former lawyer, Tom Fitch. With their help, Wyatt became an active real estate investor and prominent citizen of San Diego.

The Earps' greatest real estate investments and social success would come during these years. Wyatt was listed in the San Diego City Directory of 1887 as a "capitalist," with real estate holdings that covered two city blocks. He owned several gambling places, but spent most of his time at the Oyster Bar in the Stingaree District. He judged horses and refereed fights at the Escondido Fair. He also traveled down to nearby Tijuana, where anything too wild for San Diego ended up: day-long festivals featured cockfights, bullfights, and epic prizefights that could go seventy-five rounds or more. The Tijuana weekends attracted thousands of attendees, packing the railroad trains so full that people sometimes had to delay their return until the following Tuesday or Wednesday.

The highlight of the San Diego real estate boom was the development of Coronado Island. A special beach railroad carried prospective investors and sightseers from the ferry to the site of the construction. Josephine was there when the first lots were auctioned from a big tent, and it did indeed seem like a circus to her. She also saw the foundation of the "great rambling structure" of the Hotel del Coronado, and attended the opening gala, at which Lily Langtry performed.

Bat Masterson turned up in San Diego. He was working as a detective and recruited Wyatt to join him on a quick trip to bring back a murder suspect. Josephine joined them, and they spent a pleasant day in Ensenada, notable for a great Mexican dinner and some jewelry shopping, and soon they were all back on board the small boat to San Diego with their prisoner. It turned out to be a memorable trip because of an altercation that occurred not with the fugitive but with the captain of the ship, who demanded that Josephine and Wyatt vacate their cabin to make way for the general of the Mexican army and his staff. Wyatt refused.

Considering the men that Wyatt had killed as sole judge, jury, and executioner during the Tombstone Vendetta Ride, Josephine could not help reflecting on this unexpected showdown. She was confused and sleepless in their cabin, while Wyatt slept like a baby. Her husband had always stood on the side of the law, but this time she believed he had been in the wrong; the captain was the master of the ship, and surely his jurisdiction put everyone under an obligation to follow his orders. Wyatt defended his actions:

> *"When laws are made, Josie," he explained, "and certain powers are entrusted to a man, it is expected that the man will measure up to his trust; that he will play no favorites in the use of that power. That captain is too small for his job. He demanded that we give up our rights so that he could favor someone he wanted to impress. He was outside of his rights in that and I couldn't keep my self-respect except by refusing. If it should come before a fair court, I would be upheld by the law and he knew it or he wouldn't have let me get by with it."*

"Well, I could see his point clearly enough," she reflected, "but I still would not have had courage to defy the captain of the ship in strange waters. I'd still be willing to let the brass buttons have the right of way."

This was as close to doubt or disapproval about Wyatt as Josephine ever went. The captain with his brass buttons who was "too small for his job" sounded a lot like Johnny Behan. Wyatt's definition of justice allowed him to select those legal requirements he would obey. Self-confidence and real-time decision-making were essential to Earp ethics: decide now, and hope for a smart judge later. Josephine did not criticize Wyatt, but neither was she wholeheartedly endorsing his philosophy.

WYATT'S LAW ENFORCEMENT assignment with Bat was a temporary distraction from what Josephine loved best about San Diego: horseracing. The sport provided a substantial source of income and delight for Wyatt, plus a dangerous new passion for Josephine. "He bought only one car in his life," Josephine noted, but he never lost his childhood love of horses. She was talking about her own inclinations, as well as Wyatt's, when she observed that "this love of horseflesh, coupled with his susceptibility to the wiles of Lady Luck, formed a combination that made it almost inevitable that at some time during his career the horse-racing game should claim him." Their first racehorse, Atto Rex, was Wyatt's prize in a high-stakes poker game. His prowess as a stable owner grew from there. When his horses were successful, Wyatt enjoyed buying gifts of jewelry for Josephine: once, it was a beautiful ruby bracelet; another time, a sparkling brooch in the form of a peacock encrusted with diamonds.

Josephine liked traveling around the California racing circuit and being a welcome guest in glamorous hotels, like Hollenbeck's in Los Angeles, which attracted many of the Arizona old-timers, and Lucky Baldwin's hotel in San Francisco. She knew the name and age of every horse, and crooned to them in their stables. Wyatt sometimes rode the horses to victory himself, but as their stable grew, they hired some of the best jockeys of the day and outfitted them in the Earp racing colors—navy blue polka dots on a white field.

As they became entrenched in this new life, Josephine felt protected from the terrors of Tombstone. Horseracing was her bridge to a wonderful new social life. Unlike the boomtown gambling halls, there was a place for women at the San Diego racetrack. Josephine eagerly stepped into a more public role in this exciting tableau, wearing elegant clothes and jewelry and drinking champagne. Still the common-law wife of a celebrated gunfighter, saloonkeeper, and stable owner, she relished her new freedom and social status. Her aspirations rose still higher through her friendship with Elias Jackson Baldwin, already famous as "Lucky" Baldwin. Although considerably older than both of them, Lucky became one of the few friends who would be as close to Josephine as to Wyatt.

Born in Ohio, Lucky Baldwin came to California by wagon train. Everything he touched seemed to succeed, especially his major bet on the Comstock silver mines. With an immense fortune to invest, and an outsize ego to match, he turned from mining to real estate in the 1870s, and became one of the biggest landowners in the country, with over 63,000 acres in California alone, as well as a major owner of thoroughbred racehorses, which competed under the banner of his Rancho Santa Anita.

Lucky never swore, never drank, and never bet on his own horses, though he was always ready to stake a friend or throw a major bundle at a promising investment. He admired Wyatt—they met as two celebrity gamblers and real estate investors—but he adored Josephine. Apparently they were not lovers, though Lucky was a notorious rake who chased women as compulsively as he acquired land and horses. He was eventually married four times, and would face more than one angry challenge at the end of a gun barrel for his extramarital exploits.

Wyatt expressed less jealousy about Lucky's propensity to buy gifts for Josephine and lend her money to place bets on the horses, and more concern about Josephine's gambling debts. "Are you game enough to take our losses too?" Wyatt asked, probing this new recklessness. She knew that fortunes were up and down, Josephine protested, and her gambling was only a way to spend more time with Wyatt.

But the reality was far more troubling. Wyatt appreciated the difference between betting as a compulsion and as a livelihood. His concern seemed justified by Josephine's actions one night, when he returned to their hotel uncharacteristically drunk. Josephine waited until he fell asleep and stole his wallet, thick with the day's receipts from the saloon (including money from Wyatt's partners) and designated for a real estate investment the next day. Josephine counted its contents over and over, marveling at the first $500 bill that she had ever seen. Pretending that she was committing a crime, she filched $10. Deciding that she might as well increase her stake, she took $20, then $50, then $100, giggling to herself all the while. However, Wyatt was not too hung over the next day to forget the exact contents of his wallet. Under his rather mild questioning, which included an offer to give her an equal sum for her own purposes, Josephine confessed all and gave him back the money.

This incident ended with Josephine's promise to be more careful. Her resolution to "bet smarter" had a ring of sincerity—but it did not stick. She continued to rely on Lucky Baldwin as her banker and pawnbroker. To cover her mounting losses, she borrowed money from him, using her jewelry as security. Wyatt bailed her out and retrieved the gifts he had given her, but finally he lost patience, perhaps also concerned that Josephine was testing the limits of Lucky's valuable friendship.

Josephine submitted to a lecture—the sternest Wyatt would ever give her in their life together. "You're not a smart gambler," Wyatt told her. "You have no business risking money that way." He warned her that he would no longer redeem her jewelry.

"That woke me up," she reported in her memoir, seemingly chastised.

Her banker backed out: Lucky confirmed that he was under orders from Wyatt not to loan her any more money for betting.

JOSEPHINE RETURNED LESS often to San Francisco, and for a while, her ties to the Marcus family seemed to loosen. But her happiness was far from complete.

Josephine was furious to discover Wyatt's ties to San Diego brothels like "the Golden Poppy," where prostitutes dressed in colors that matched the paint of their respective rooms, just a few steps upstairs from his successful Oyster Bar and Gambling Hall. Wyatt's absences from San Diego grew more frequent when he opened a saloon in Harqua Hala, Arizona, an emerging boomtown, where he drew openly admiring glances from women wherever he went. He also returned to Colton, California, at least once to see his parents, a nine-hour train ride on the recently completed Southern Pacific line that had connected San Diego to the rest of California.

While he was away, Josephine would open his mail, ostensibly to answer his fans, but also to "nip any unwelcome contact in the bud." She knew just the type of good-looking, intelligent women that he liked. The fact that he was not her legal spouse did not inhibit her from acting like a jealous wife. "While I was proud when women noticed my husband," she declared, "I realized that he was human and not all of the attentions with which they showered him were sprung from unmixed motives." They were nearing the ten-year mark of their relationship, and had designed a "little custom" that she considered key to domestic peace. When she lost her temper, which was often, Wyatt just grinned until her good humor was restored. But if her fury crossed some threshold of intensity, he would take his pipe and hat and walk out. She then wrote down the rest of her angry speech and left it for Wyatt to find. He called the notes her "love letters," which gave her the privilege of the "last word."

Is this young beauty Josephine Sarah Marcus?

Possible age progression for Josephine, based on forensic
analysis. Only the last three photographs of older Josephine
are fully authenticated.

*Below left:* Celia "Mattie" Blaylock, Wyatt's common-law wife
before Josephine

*Below center:* Wyatt Earp, mid-1870s

*Below right:* Johnny Behan, Josephine's lover and Wyatt's rival

O. K. CORRAL
Livery and Feed
Stable.
Allen Street, between Third and Fourth.
TRANSIENT STOCK WELL CARED FOR

Josephine and Wyatt attended the Chicago
Exposition and Buffalo Bill show in 1893.

Henrietta Marcus Lehnhardt,
Josephine's younger sister

Artist Edna Lehnhardt, Josephine's niece

## EARTHQUAKES, EXPOSITION, AND SCANDAL

Emil Lehnhardt, "Candy King of Oakland," bought this mansion on Telegraph Avenue as a wedding gift for Henrietta. Josephine and Wyatt lived there after the 1906 earthquake.

Wyatt Earp's saloon in Tonopah, Nevada, circa 1902.
Josephine may be the woman on the left.

Nome beach scene, summer of 1900

THE GREATEST MINING CAMP
THE WORLD HAS EVER KNOWN

Stuart Lake

HELLDORADO AND
EARPMANIA IN THE 1920S

# WYATT EARP
## *Frontier Marshal*
## STUART N. LAKE

# WAITING FOR WYATT

Wyatt and Josephine and dog Earpie at the "Happy Days" mining camp in Vidal, California, around 1920

# WESTERN UNION

| CLASS OF SERVICE | | SIGNS |
|---|---|---|
| This is a full-rate Telegram or Cablegram unless its deferred character is indicated by a suitable sign above or preceding the address. | | DL = Day Letter<br>NM = Night Message<br>NL = Night Letter<br>LCO = Deferred Cable<br>CLT = Cable Letter<br>WLT = Week-End Letter |

NEWCOMB CARLTON, PRESIDENT    J. C. WILLEVER, FIRST VICE-PRESIDENT

The filing time as shown in the date line on full-rate telegrams and day letters, and the time of receipt at destination as shown on all messages, is STANDARD TIME.

**Received at 341 Plaza, San Diego, Calif.** Always Open

SB364 23=LOSANGELES CALIF 15 1228P

STUART N LAKE=                                    1029 JAN 15  PM 12 34

            3916 PORTOLA PLACE SANDIEGO CALIF=

FUNERAL WEDNESDAY MORNING TEN OCLOCK PIERCE BROTHERS SEVEN

HUNDRED TWENTY WEST WASHINGTON STREET WILL EXPECT YOU AND MRS LAKE

BE SURE TO COME=

            JOSEPHINE S EARP.

THE QUICKEST, SUREST AND SAFEST WAY TO SEND MONEY IS BY TELEGRAPH OR CABLE

Pallbearers at Wyatt's funeral. *Left to right:* W. J. Hunsaker, George Parsons, John Clum, William S. Hart, Wilson Mizner, and Tom Mix.

Josephine and Vinnolia Ackerman
(*left*) in 1939

Josephine (*left*) with Rae, Mabel, Ernest, and Jeanne Cason

Photo of *Wyatt Earp* inscribed to "Mrs. Wyatt Earp from Lincoln Ellsworth, 25 February 1934."

**EARP WIDOW HONORS EXPLORER**

Mrs. Josephine Earp, widow of Wyatt Earp, mining man and peace officer of Tombstone, Ariz., yesterday presented an autographed copy of the book, "Wyatt Earp" to Lincoln Ellsworth, Arctic explorer. Ellsworth named his exploration ship, Wyatt Earp, in memory of the man he calls his boyhood hero. The ship is now in dry dock at Dunedin, N. Z., undergoing minor rudder repairs. In addition to the book, which is a story of the winning of the West as reflected by the life of her late husband and on which she collaborated, Mrs. Earp gave the explorer an autographed photograph of the peace officer. The photo, one of the last made of Earp, is to fill a prominent nook in the cabin of the exploration ship, along with other mementos of the hazardous days of Wyatt Earp. The picture shows Mrs. Earp and Ellsworth.

Josephine greets Lincoln Ellsworth upon his return from Antartica aboard the *Wyatt Earp*.

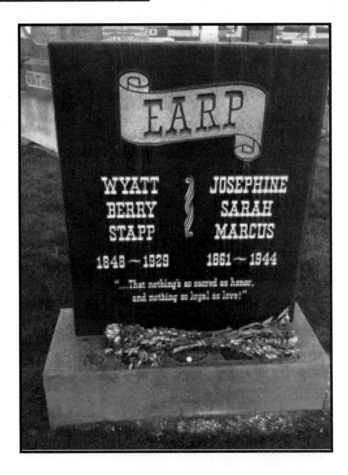

Flood recorded Josephine's angry tirades delivered through his locked door: *"I'll get back at you – good and hard."* Mrs. Earp Sunday PM—April 18, 1943.

It must have worked, because when it was time to leave San Diego, they were a united couple again. Their departure was not because of marital problems but the result of a major collapse of the real estate market. San Diego had its boom, which was followed by the inevitable bust. Wyatt's substantial holdings in the Stingaree district were affected first. As the general business climate began to sour, moral outrage over the libertine atmosphere grew and the clamor for reform became a convenient political issue that temporarily obscured the dangers inherent in San Diego's overvalued assets. "The growth of the evil has gone on through the sufferance of the authorities and it is high time the law was enforced," thundered the *San Diego Union*. The police chief was indicted, and Tombstone friend William Hunsaker was accused of representing criminals and gamblers. Fees for saloon licenses tripled. Gambling rooms and dance halls were shut down. Real estate plummeted in value, and the banks pulled back loans. Businesses shuttered, construction ceased, and the city began to lose population. Alonzo Horton, who had been responsible for so much of the early excitement, was wiped out.

By 1890 the boom was over. Although the city would eventually recover and thrive again, it was time for Josephine and Wyatt to move on, leaving behind loose ends in the form of lingering lawsuits for recovery of promissory notes and real estate taxes.

They had come so close to making their fortune. The scene of her greatest triumph, San Diego could have provided the permanent home and the social prominence Josephine craved. However, as the once bright future slipped away, she had no regrets. They were leaving together, and memories of their glamorous horseracing days would outlast the disappointment of their now worthless real estate holdings, her gambling losses, the seedier side of the saloons with the whores upstairs, and their domestic quarrels. She would always remember these years as a beautiful and colorful "kaleidoscopic picture":

*I have but to shut my eyes and call them to mind to see again the gay sights and hear the merry sounds of those full years—Santa Rosa, Chicago, Tanforan, Laton, Cincinnati, St. Louis, Kansas City, Santa Anita . . . crowds screaming, yelling, roaring; bright banners; bold trumpets, gay parties, lovely women, handsome men, special trains, horses, smells, music, brave laughter—and Wyatt coming home to me; if he had won, the light of triumph in his eyes; if he had lost, the gleam of hope for a better showing next time.*

Some four hundred miles east of San Diego, Mattie Blaylock Earp was confronting the bleak reality of life with no glamour, hope, or love. She left Tombstone bearing the name of Mrs. Earp and endured a few excruciating months at the home of Nicholas and Virginia Earp until she could no longer ignore the knowing glances that passed between his parents and brothers. Wyatt had moved on. In the early days of their separation, he had sent her money, but by 1888 she was broke and alone. Her family had lost contact with "Ceely," as they called her. In despair, she spiraled down into a grueling cycle of poverty and prostitution. She was living among strangers, eager to tell anyone who would listen that Wyatt Earp had "wrecked her life by deserting her and she didn't want to live."

On the morning of July 3, 1888, in the small town of Pinal, Arizona, Mattie asked the man in her bedroom to buy her another bottle and another vial of laudanum. She took a swig of whiskey with the man, and then he left her with the opiate drops. By the time her next customer arrived, she was dead.

It would be decades before anyone connected Mattie Blaylock with Wyatt Earp. This was the secret that Josephine most dreaded, that the world would know the sordid details of Mattie's death and would add her to Wyatt's list of victims, leaving Josephine reviled as the cruel, adulterous accessory to suicide.

Josephine's terrified looks over her shoulder would be the driving force behind many of her future actions. Nothing less than Wyatt's legacy, as well as her own reputation, was at stake.

The echoes from Tombstone had been muted in San Diego. They were back again now, louder than ever.

As if she and Wyatt had been waiting for Mattie to release Wyatt of any perceived obligation, Josephine later claimed that they were married soon after Mattie's death, legally wed in a ceremony on Lucky Baldwin's yacht. The year was 1888.

There is no evidence that Baldwin ever owned a yacht, though his daughter did. Nor have any records of an actual marriage ceremony ever turned up.

Nobody believed the story of a legal wedding aboard Baldwin's yacht as anything other but Josephine's wishful thinking. Josephine's biographers tried to corroborate her story, and when they could not, they sought to reassure her that it did not matter. But the question of her legal status mattered very much to Josephine, as did Mattie's suicide.

For the rest of her life, she struggled to keep these two secrets: her supposed wedding on Lucky Baldwin's yacht and Mattie's last days as Mrs. Earp.

AFTER THEIR BUSINESS reversals in San Diego, Josephine and Wyatt regrouped in San Francisco; it was the city they loved most, and an important time for Josephine to be closer to her family. Her father had fallen ill and died on January 5, 1895, the first of the Marcus family to be buried in the Jewish Hills of Eternity cemetery. Mrs. Marcus moved to the Lehnhardts' big house in Oakland. The Earps lived first with Josephine's older sister Rebecca and her husband Aaron at McAllister Street. Wyatt renewed old acquaintances and began to build up a new stable and betting connections. When his horses won, Wyatt would come home and spill out a bag of twenty-dollar gold pieces on his sister-in-law's kitchen table. Luck was running in their favor more often than not, and soon they moved to a place of their own in the old Bay District near Fulton and Golden Gate Park, with a stable big enough for six horses. No longer a "capitalist," as he had been in the San Diego directory, Wyatt was now officially listed as a "horseman" in the 1896 San Francisco directory.

Josephine's favorite horse was Lottie Mills, a smart, beautiful thoroughbred that Wyatt bought in San Francisco and owned with Lou Rickabaugh, his former partner back in Tombstone. Lottie Mills attracted much attention whenever she competed, and even Lucky Baldwin wanted to buy her. Wyatt had other plans.

Chicago's 1893 World's Columbian Exposition was a watershed event in America history. It drew people from all around the world, including Josephine and Wyatt. Josephine was excited to plan their trip, which would include racing Lottie Mills. On the way east they stopped in Colorado, where the *Denver Republican* ignored Josephine but gushed over Wyatt's suave appearance with the headline "He Is a Dude Now," noting approvingly his "neat gray tailor-made suit, immaculate linen and fashionable neckwear. With a derby hat and a pair of tan shoes he was a figure to catch a lady's eye and to make the companions of his old, wild days at Tombstone and Dodge, who died with their boots on and their jeans pants tucked down in them, turn in their graves."

The Exposition was a celebration of the four hundredth anniversary of Columbus's arrival in the New World. Opened officially by President Grover Cleveland, it attracted some 6.8 million guests, a veritable *Who's Who* that included Lucky Baldwin, Mark Twain, and many European heads of state. Many visitors would consider the fair one of the highlights of their lives. The White City, so called for the bright stucco of its buildings and the glare of electric streetlights, was the inspiration for L. Frank Baum's depiction of Oz. Josephine could have amused herself on the world's first Ferris wheel, which offered remarkable views of the Exposition and its host city. She could have visited a reproduction of Marie Antoinette's bedroom, or re-creations of a Cairo street, an Eskimo village, a Moorish palace with $1 million in gold coins, or a full-size model of the Parthenon. Every state of the union and many countries constructed pavilions and displays; the Arizona Territory exhibit featured Indian history as well as the region's natural resources.

But history, technology, and the wonders of the global village were mostly lost on Josephine. She was most excited about horseracing and the chance to meet Buffalo Bill, Wyatt's acquaintance from the days of buffalo hunting. Lottie Mills was victorious at the track; almost as thrilling was the run of an unknown horse named Boundless who came from behind to win the American Derby and a $50,000 prize.

Had she been so inclined, Josephine could have attended the nearby annual conference of the American Historical Association, held in conjunction with the Exposition. She might have paused to listen to the young historian Frederick Jackson Turner deliver one of the most provocative and influential lectures of its kind: "The Significance of the Frontier in American History." Based on the results of the 1890 census, Turner argued that the frontier was "closed" because there were no longer any great tracts of land to settle. "The frontier has gone, and with its going has closed the first period of American history," Turner pronounced. He evoked the ideals of the vanishing frontier as the greatest qualities of America: democracy, self-sufficiency, rugged individualism, a practical outlook on life, and "master grasp of Material things."

As if to underscore Turner's point with head-spinning irony, the highlight of the Exposition for Josephine and Wyatt Earp was to be among the six million visitors who saw Buffalo Bill's reenactment of the western adventures they had so recently experienced, a hint of what they would experience when Hollywood invaded Tombstone.

Soon after Turner's lecture, *Harper's Monthly Magazine* commissioned "short stories of Western life which is now rapidly disappearing with the progress of civilization." Their writer, Owen Wister, went on to write one of the most popular novels of cowboy life, *The Virginian*, the best-selling book of 1902 and 1903.

"[The West] is a vanished world," Wister wrote. "No journeys, save those which memory can take, will bring you to it now."

Underlying Turner's lecture and Wister's fiction was the question of how America would adapt to the passing of the frontier. The same challenge would resonate through the rest of Josephine's life: Could she and Wyatt survive in postfrontier America?

## 3 | THE GREATEST MINING CAMP
## THE WORLD HAS EVER KNOWN

*S*HE WAS no longer the runaway teenager throwing it all away for love. Josephine had spent more than ten years as Mrs. Wyatt Earp, in boom-towns and mining camps in Colorado, Texas, California, and Arizona. There was always a flurry of public recognition when they entered a new place, and Josephine listened cautiously for whispers about Mattie or the validity of her claim to be Wyatt's wife. She heard none, other than the low hum of fear she generated herself, a constant dread of discovery and scandal.

In the summer of 1897, Josephine and Wyatt were living in the desert outside of Yuma, Arizona, when news of the Alaskan Klondike gold fields hit. Even in Yuma, Josephine heard fabulous stories of lives transformed by untold riches. Soon the town was so full of gold rush mania that the editor of the local *Sun* volunteered to build a two-hundred-passenger ship to deliver fellow Yumans directly to the gold fields. With mock seriousness, he proposed to hire local notables as journalists, and he nominated Yuma's own Wyatt Earp as his law and order correspondent. Then suddenly old comrades from the boom days of Tombstone, San Diego, and San Francisco began to write to Josephine and Wyatt with one message: Come to Alaska!

Josephine badly needed a new escape plan. Just a few months before, the Earps fled San Francisco in the wake of a scandal that came not from any revelations about Mattie Blaylock's suicide but from an accusation that Wyatt had fixed a high-stakes prizefight. The trouble started with a sensationalized three-part account of Wyatt's frontier adventures. The *San Francisco Examiner* articles sold newspapers, but the exaggerated accounts of Wyatt's exploits set him up for ridicule. On the heels of that misstep, Wyatt agreed to serve as a referee for a championship prizefight between Tom Sharkey and Bob Fitzsimmons. Fitzsimmons was heavily favored to win the $10,000 purse. The promoters had quarreled over the choice of referee, and compromised on Wyatt only a few hours before the fight.

Josephine and Wyatt dined that night at Goodfellow's Grotto Café, not far from Mechanics' Pavilion, where a capacity crowd of 15,000 was already gathering, one of the first to include women. Josephine had no love of the ring and went home after dinner. Her nephew Isidore, who was interested in anything that interested Wyatt, accompanied him.

Wyatt entered the ring wearing a grin—and a gun. He made no attempt to conceal the weapon, which was tucked into his belt and visible to everyone after he removed his coat. Just darned forgot he was wearing it, he protested when the ringside audience gasped. Decades after Tombstone, where a frontier lawman worked, gambled, and slept with a pistol, Wyatt was "no more conscious of its being there than of my coat or my vest," he tried to explain to a shocked police officer. Wyatt surrendered the gun without hesitation. But the crowd was already riled up.

The fight began. In round eight, Wyatt called a foul against Fitzsimmons and awarded the decision to the underdog, Sharkey. Wyatt was nearly crushed in the chaos that followed his unpopular ruling. He was defended by Lucky Baldwin and other ringside observers—and by some of the doctors who examined Sharkey—but all of that was about to be buried in an avalanche of negative press.

The next day's headlines screamed "Fitzsimmons Was Robbed!" Sharkey quickly cashed his check for $10,000 and left town. From their rooms at the Baldwin Hotel, Josephine railed loudly against what she considered to be personal attacks against Wyatt, orchestrated by the owner of the *San Francisco Chronicle*, who happened to be one of the big losers that night. The newspapers kept the story alive for weeks, sometimes illustrated with mocking cartoons of Wyatt as a corrupt frontier lawman in the ring, pistols at the ready.

A decade of respectability and careful work to silence the echoes of Tombstone collapsed. The facts were ambiguous. But what was indisputable was that the Sharkey-Fitzsimmons fight put all the stories about Tombstone, all the lingering tales of Wyatt as a cold-blooded killer and bad man, back on page one, newly amplified with juicy accusations of theft and corruption. "The old lies come bobbing to the surface of the ocean of printed matter," Josephine lamented. It was not Mattie who had returned to haunt her—at least not yet—but the humiliation was particularly hard to bear in San Francisco, home to Josephine's family, including her socially prominent sister Hattie and her upright brother-in-law Emil Lehnhardt.

In the aftermath of the fight, Josephine was desperate to escape the finger-pointing and notoriety. Ably represented by a Wells Fargo lawyer, Wyatt appeared at police headquarters to answer a charge of carrying a concealed weapon. He paid his fifty-dollar fine, but that did nothing to quell the public outcry. Proclaiming herself tired of the "contention and greed" of the sporting world, Josephine was relieved when Wyatt proposed a permanent move to the California desert to raise cattle or to prospect around the Colorado River.

This would be Josephine's first camping experience. A city girl, she instinctively feared the desert and its barren isolation. But time alone with Wyatt was always welcome, and she waged an internal battle to conquer her fears. "Notwithstanding my faint courage in the face of danger, I had a certain spirit of adventure that lured me into the very experiences that I feared most. I knew, however, that Wyatt had enough courage for both of us and felt that as long as he was with me I would dare anything." She had no home, no crops or store to tend, no children to educate, no compelling ties to any place in the world.

Hoping that their infamy and embarrassment would dissipate in the "sweet breath of the desert," the couple sold off some of their investments. Wyatt happily slid back into the outdoor life, "like a boy in his eager enjoyment." To Josephine's delight, their retreat awakened in her a love of nature, the more rugged and unusual the better, and a particular affinity for the desert.

The respite would be brief, hardly the prolonged retreat that Josephine had sought. The gold rush chatter proved irresistible to Wyatt, who proposed that they return immediately to San Francisco and outfit themselves for Alaska. Josephine was reluctant to sell off the camp gear that they had so recently acquired and to return to the city that had just been the site of their greatest disgrace. But Wyatt's warning about Alaska—"There will be hardships"—had the reverse effect of overcoming her fears. She still thought of herself as a "scaredy-cat" but was proud that she had survived in a dozen exotic locales. "Hardships, hmph!" she responded to Wyatt. "I'd like to see the hardships a man can endure that are too great for a healthy woman. They will not be any harder for me than for you." Her next words coyly evoked the conventional marriage vows that she had not yet spoken, and never would: "Besides, I promised to stick with you 'till death do us part' didn't I?"

Her Alaska sojourn would turn out to be the most demanding time of her life, but also the happiest. Surely there, so close to the Arctic Circle, she could lay down the burden of the past.

. . . .

"ON THE TRAIN to San Francisco all we could hear was GOLD and the KLONDIKE," Josephine recalled. Alaska "was a remote frozen country few of us had heard of before and which seemed as far away and unknown as the land of Cathay did to Europeans in the days of Marco Polo." Their plan was to make their way to Dawson, the capital of the Canadian territory, via the recently opened Chilkoot "White" Pass.

An accident delayed their departure and changed their lives. Wyatt was riding down Market Street on a public streetcar when he caught sight of a crowd standing around the first automobile that most of them (including Wyatt) had ever seen. Hoping for a closer look at the shiny object drawing all eyes, Wyatt jumped off the trolley, slipped on the rain-slick street, and bruised his hip badly enough to keep him in bed for three weeks. Steamer after steamer left for Alaska. Finally Wyatt pronounced himself well enough to book passage for two on the *City of Seattle*, bound for Wrangell, one of the popular gateways to the Klondike.

Josephine and Wyatt entered Alaska at the moment when the wilderness was being pushed back forever. Just twenty years before, naturalist John Muir had declared Wrangell "the most inhospitable place at first sight I had ever seen." That Wrangell was gone; in its place was a rough-and-tumble mining town like so many other boomtowns. This one was just that much more remote, and soon, much colder. Despite the obvious climatic differences between the Arizona desert and the Alaskan frontier, northern boomtowns were equally lively with the familiar sounds of saloons and gambling. Fortunes were being taken out of the ground, but also out of the pockets of the miners. Con artists proliferated, including one former Tombstone acquaintance, Jefferson Randolph "Soapy" Smith, who reigned briefly as chief swindler of boomtown Skagway, before being killed in a gunfight. Among Alaska's crop of unusual entrepreneurs was Josephine's friend Sidney Grauman. The future movie impresario of Los Angeles got his start selling fifty-dollar newspapers to Alaskan storekeepers, who read the news for a fee to homesick miners.

Josephine and Wyatt inquired about getting to the gold fields and the northern prospecting towns. The fastest route would take them over the Chilkoot Pass, later made famous by Jack London as "the worst trail this side of hell," thirty-three miles of treacherous hiking where thousands of would-be fortune hunters struggled like ants up nearly forty-degree inclines. Once they descended, conditions were hardly better on the other side. Miners endured a life of hunger, fever, dysentery, and scurvy. Had Wyatt's accident not delayed them, Josephine might have taken her turn on the so-called Golden Staircase. But the time was dangerously close to the onset of winter, and Josephine feared that Wyatt was not yet well enough to endure the arduous mountain crossing.

In the end, it was not only cold weather and Wyatt's bruised hip that changed their plans: Josephine discovered that she was pregnant.

She had been pregnant at least once before, and may have undergone an abortion in Tombstone or in San Diego, according to dark hints that she dropped to family members and friends. But now Josephine wanted to be a mother. At thirty-six, she was relatively old to be bearing her first child. She was confident that Wyatt would propose an immediate return to the States. He did: Wyatt was as eager to have a child as she was.

Josephine's pregnancy ended in a miscarriage. The stillborn child was a boy, although there is no record of an infant's burial. According to conversations reported by family members and friends, the loss was devastating. Josephine's age, their nomadic life—all were poor indicators that she would have another chance. She blamed her own anatomy for the loss of her baby, lamenting, "God didn't make me right for having children." But sorrow and disappointment did not curdle into bitterness. In later life, she delighted in the pregnancies of friends, and was quick with a gift or telegram to share the joy of a new baby. She was playful and affectionate with young children, whether they were strangers or her own nieces and nephews. Albert Behan remained a close friend, an improbable relationship that survived the bitter breakup between Josephine and his father, as well as the still more divisive events that pitted Wyatt and Behan against each other at the O.K. Corral and during the Vendetta Ride.

Her miscarriage slowed her down but did not end gold rush dreams. "We had to try Alaska again," Josephine recalled. She recuperated in San Francisco. As soon as spring weather arrived, she declared herself fit for another run at Dawson. However, conditions at the Chilkoot Pass had grown still more dangerous. The pass had recently been the site of a massive avalanche in which about sixty people were buried alive. This time, Josephine and Wyatt chose the slower but safer all-water route, which would take them up the west coast to Alaska, then along the Yukon River to St. Michael, a small settlement near the mouth of the Yukon, and then to Dawson.

They negotiated to get tickets on "any old tub that could be got to float" and found space on the creaky SS *Brixon*. After a few days, uncomfortable conditions and meager rations caused a passenger rebellion, which escalated into a full mutiny of the crew. Josephine was proud that Wyatt was asked to arbitrate the dispute, after which the captain regained control in time to fill the *Brixon*'s food stores at the port of Unalaska. They reached St. Michael, expecting to transfer to another vessel, but found there only a half-built skeleton of a boat. Delays multiplied: after construction was finished and the boat declared seaworthy, the newly christened *Pingree* sprung a leak on its maiden voyage and limped back to shore. Nervous about the approach of winter, the passengers passed the time cutting wood for fuel in the mosquito-thick forests along the riverbank. Finally, the boat was ready to begin its 1,500-mile voyage through the winding Yukon River, where shifting currents and dangerous sandbars challenged even the most experienced captains.

As she had fallen in love with the California desert, now Josephine reveled in the "wildly picturesque" landscape of the Yukon River, flanked on both sides by steep mountains with tall trees marching down to the water's edge. The wind was still warm with the last summer breezes. But soon Josephine felt the autumn chill and could see the shallow parts of the river beginning to freeze. Snow was falling steadily as they drew close to Rampart, a small settlement halfway between St. Michael and Dawson. The steamer struggled forward as more ice solidified. One more bend of the river—and they were locked in a deep freeze.

The passengers left the boat. Some of them made immediate plans to continue their trip by dogsled. Josephine canvassed the opinion of acquaintances, who predicted that business would be better in the American part of Alaska. "You are under the good old Stars and Stripes," they urged. "Up at Dawson, the Union Jack flies above you and they don't favor Yankees."

Rampart would be her home that winter.

THE SMALL VILLAGE was raw and new. Originally established by settlers along the Yukon River, Rampart had become a hub for prospectors and trappers and home to about two thousand men and fifty women. It was part of the rapidly growing empire established by the Alaska Commercial Company, a Russian trading company that had been chartered by Catherine the Great, then sold to a firm of San Franciscan Jewish businessmen who built it into a major provider of general merchandise for trappers, explorers, and gold seekers, steamboat transportation, and financing of Alaskan mining ventures. The company's village stores became the multipurpose center of community activities, serving as bank, post office, and the venue for marriages and funerals. Rampart had no church, though Josephine had heard about one minister in town "looking for business."

Rampart even had a connection to the Arizona Territory through Tombstone founder Ed Schieffelin, who had moved to Rampart after selling his silver mines. Schieffelin believed that Alaska was the best place to tap a mineral belt that encircled the earth, and he invested in a steamboat from the Alaska Commercial Company. But he was deterred by the inhospitable climate and sold his riverboat to the unofficial mayor of Rampart, Captain Al Mayo, and returned to the States.

Rampart society was tightly knit but welcoming to newcomers. Although it had none of the sophistication of San Diego or San Francisco, it was the first place that Josephine had ever lived where a gambler and his common-law wife could socialize freely and enjoy the company of a diverse and remarkable cast of men and women. Instead of spending most of her time waiting for Wyatt, as she usually did, they were together day and night. Their closest friends were Captain Mayo, a former circus acrobat who had been among the first traders in the Yukon Valley, and his wife Maggie, the daughter of the chief of a nearby Indian trading village. Intelligent and resourceful, Maggie was her husband's trusted business partner, and also served as translator and expert seamstress. Josephine observed with interest how Maggie skillfully navigated her two worlds, keeping house "in the manner of a typical American housewife," giving her children "every advantage," but also maintaining strong ties to her native family.

It was Maggie who tutored Josephine in the art of keeping house in the harsh Alaskan winter. Josephine embraced this unfamiliar world, a new environment that she was determined to conquer. She was soon outfitted with a beautiful parka of white deerskin, made by Maggie, while Wyatt purchased his cold-weather gear from the local Indians. Finding a suitable residence was tricky, since Rampart had fewer than five hundred cabins. Some less fortunate residents were forced into unusual arrangements, like one woman who shared her place with a roommate who worked in a bar: it was her "rent" to keep the cabin warm during the night, and when her roommate returned to sleep during the day, she went to work. Josephine found a one-room cabin for a monthly rent of $100. Close to a brook, now ice-locked and silent, the cabin was made of "logs chinked with moss and a roof covered with a deep layer of earth," furnished with crude home-made furniture. Josephine carpeted the floor with overlapping burlap sacks obtained from local storekeepers, layered with furs that had been traded from Indians or Wyatt's own hunting. The bed was warmed with blankets and down comforters that she brought from California. Calico curtains hung on the windows and separated the bedroom from a small sitting area. They kept the fire going day and night, grateful for the plentiful supply of wood in Rampart.

Josephine improvised a cozy home, the only one that she would ever recollect with nostalgia and pride. The resulting scene of domestic bliss stood out among the long list of hotels and campsites that they called home. "As soon as we were all settled and I was taking my first batch of bread from the oven, with beans boiling on the stove, Wyatt came in from battling a snow storm, dressed in his mukluks [boots] and parka. He sniffed, his eyes lighted with pleasure: 'Snug as a bug in a rug!' he exclaimed. On such small hinges does the door to contentment swing."

Rampart was the first kitchen where Josephine cooked regularly and presided over dinner parties. She apparently brought no recipes from her mother's Jewish kitchen, preferring to prepare Wyatt's favorites, mostly southern specialties such as hot biscuits with the occasional luxury of butter and sweet preserves. In the desert, she and Wyatt had shared the cooking and cleaning duties, but here she became an accomplished cook and popular hostess who was known for her ability to coax some degree of novelty into her menus within the extreme limitations of a Rampart larder. Milk and butter were scarce, and eggs even more rare, priced as high as three dollars per dozen. Moose and caribou were plentiful, and she paired them with beans and potatoes. Ptarmigan, a local bird, passed for chicken. She had the only supply of garlic and chili in town, packed among the provisions she had brought from San Francisco. Her fresh ice cream was a local favorite, a simple concoction of sweetened condensed milk (she recommended Eagle brand), powdered eggs, and a little vanilla, and she might have been stifling a laugh when she instructed that the bowl could be "set out of doors where it would freeze almost immediately." Rampart cooks had to have a sense of humor; one of Josephine's elegant meals ended with attractive slices of cake made out of a large white life-preserver, slathered with real icing.

Prospecting for gold proved to be far less amusing. Wyatt managed to stake some claims, but the frozen ground was like solid cement. Although one million dollars in gold had come out of Rampart the previous summer, the Earp claims had yielded nothing so far. Instead, Wyatt worked in a saloon where the daily take went as high as $700, more than enough to secure whatever luxuries were available in a snowbound Alaskan mining camp.

That winter, Josephine cheerfully counted her blessings. Given the isolation and deprivation that crippled many Rampart residents, she and Wyatt were comparatively well off. She had a new appreciation for the restorative power of a home-cooked meal, reserving special pity for prospectors who returned alone to their cabins after a day of frigid drudgery, chilled to the bone and too exhausted to cook. In snowbound Alaska, this was not a simple matter of enjoying companionship and nutritious food but a serious matter of survival. Depression and failure, compounded by hunger and loneliness, was a dangerous psychological brew. She decried the common practice of sending young men to Alaska to sow their wild oats, where they were "drawn into the maelstrom of drunkenness and gambling that existed in most of the camps," she noted with intuitive concern. "The nostalgia, the excitement, the too sudden acquisition of wealth, or the disappointment of their not finding it, the lack of a comfortable place outside of the gambling halls, dance halls, and saloon; lack of good food— any or all of these, combined with the almost universal drinking of liquor, brought many a man to a level he would never have dreamed himself capable of descending to."

Even seasoned prospectors were not prepared for the severity of the Alaskan winter, its long dark days, and temperatures that hovered some fifty to sixty degrees below zero. Suicide rates were high, alcoholism rampant. The entire town of Rampart grieved as one family over the tragedy of a California widower who sold his worldly possessions and left two small children with relatives to seek his fortune in Alaska. He was not pursuing the risky fantasy of a gold strike: his plan was to open a restaurant in Dawson. He invested his entire savings in provisions, and his greatest treasure was 1,500 dozen precious eggs, packed in lard and shipped in large barrels to Alaska. But he, like Josephine and Wyatt, was unexpectedly stuck in tiny Rampart for the winter. The eggs spoiled before he could sell them. He was found hanging from the rafters in his cabin. "I believe a wife would have helped him find a way out of his trouble," Josephine declared. She joined her neighbors in raising a fund for his orphans.

DECADES LATER, RECALLING that winter from the warmth of her California home, even the relentless cold could not mar Josephine's memories, which were forever colored by the beauty of the landscape lit by the northern lights, and the comfort of the closest female friendships she had yet known outside of her sisters. "If Rampart had not been such a sociable little community we might have found the dark and the isolation unendurable," she recalled. Doors were left unlocked. Neighbors helped neighbors. Wyatt read his way through the extensive library of their neighbor Erasmus Brainerd, a prominent resident of Seattle who was there with two colleagues to assess the possibility of promoting their city as the gateway to the Yukon.

Josephine spent long contented days and evenings with friends, embroidering, reading aloud, and swapping recipes. "I do not remember that we played cards often," she notes, with perhaps some defensiveness. "It was later at Nome that I learned to play cards and found them really fascinating." Even among the domestic pleasures of Rampart, gambling was never far away from their life, whether it was a friendly game between Wyatt and the neighbors or a high-stakes table at one of the local saloons.

The fast crowd they ran with in San Diego and San Francisco had been replaced by a different social order, cruder in its clothes and range of entertainment, yet richer in ideas and energy, free of the snobbery and ostracism that Josephine dreaded. Even the newspapers seemed to cooperate, with one San Francisco article complimenting Wyatt Earp as "a model citizen of Rampart." Josephine participated in community gatherings that ranged from lectures on the Spanish-American War in Cuba, literary discussions on books from Mr. Brainerd's collection, and sleigh rides along the frozen river behind "Napoleon," the only horse in Rampart. Tombstone, the Sharkey-Fitzsimmons debacle, even the sadness of her miscarriage—all felt wonderfully distant.

The frequent town dances were Josephine's special delight. Everyone was invited, and dance they did, to music provided by the violin of the postmaster Mr. Fleishman and the banjo accompaniment of the future novelist Rex Beach, whom Josephine described as "a tall good looking young man . . . fresh from college, with a flow of good humor that enlivened many a long winter evening." The Rampart events were known as "cookie dances," but the refreshments also included wine. The inhospitable climate and smallness of the town loosened everyone up. Their bulky cold-weather clothes were a source of amusement, as well as a great equalizer. Josephine mockingly compared their exuberant and casual parties to a formal cotillion: "Have you now a picture in your mind of several couples with powdered wigs, the men in velvet coats and satin breeches, the women in full-hooped and panniered gowns, moving through the stately measures of a minuet with courtly grace to the accompaniment of violins and harpsichord? Then banish it! Put in its place one of the strong men in mackinaws, corduroys and mukluks, and fair ladies in corduroy jackets, short skirts and—yes mukluks —but moving through the stately measure of the dance with courtly grace to the accompaniment of a violin and a banjo!"

To Josephine's surprise, Wyatt sometimes enjoyed a little more wine with his cookies than she considered good for him. While he didn't totally disgrace himself, Josephine observed a "remarkable lack of skill" in his dancing, though her criticism was perhaps sharper because he was dancing with another woman. Her favorite partner was her new friend Tex Rickard, a former town marshal from Texas and a veteran of several Alaskan winters.

Rampart lacked only fresh news and the familiar faces of her family. When Rex Beach and the local postmaster "mushed" down 850 miles from Dawson to Rampart with mail, Josephine was grateful to pay a dollar for every letter, ridiculing two young men from the East who asserted their rights as American citizens to free mail service. Writing some thirty years later, Josephine—never one to save papers or material possessions—disclosed that she still treasured a letter from her mother that came "with that [Rampart] mail and for which I would gladly have paid much more than the fee of one dollar. It is dated at San Francisco, September 8, 1898."

For her first Rampart Thanksgiving, Josephine stretched her chef's ingenuity to the maximum and produced an astonishing variety of roasted and canned meats, vegetables, and fruits, and a dessert course of fresh doughnuts, pies, and cakes. That night one of their guests was too ill to attend the dinner, and Josephine and a friend set out to deliver his meal. They were better cooks than navigators. When a sudden blizzard overtook them, they were soon lost and huddled in a thicket, cold and terrified of wild animals. Josephine was struck with a crippling moment of fear, which she associated with Wyatt's absence. "Never a brave woman, I had always depended completely upon Wyatt and now I was in a panic." Wyatt soon showed up with a small rescue party, but the memory was a powerful one.

Josephine's self-styled "scaredy-cat" image went together with a morbid fear of sickness, even when it wasn't her own life in danger. One such moment struck her when Captain Mayo asked for help in delivering Maggie's third child. Despite her close friendship with Maggie, Josephine refused. Perhaps her own recent miscarriage exacerbated her fears, but Josephine's misgivings went beyond her lack of training as a midwife. "The issues of life and death have always appalled me," she confessed. She gave in only when Wyatt sternly ordered her to go. Even after Maggie's baby was born, healthy and strong, Josephine predicted that she would be just as paralyzed with fear the next time she was needed.

Weather was everything in Alaska. Josephine learned to recognize the groaning of the ice as the signal of the spring thaw, and as the brook behind their cabin came to life with a rush of running water, she prepared for sunny days and nights with the construction of an "Arctic roof garden" on the earthen roof, a common feature that took advantage of the brief but intense growing season. With the excess sentimentality to which she was susceptible, Josephine inserted into the scene the incongruous figure of Wyatt Earp, Alaskan farmer: "The lengthening visits of the sun warmed our little rooftop garden and it was not long until the young lettuce, radish and onion plants were crowning our domicile with a jaunty bonnet of green." She compared Wyatt to a mother in his care for his young plants, "gazing rapturously at his garden."

The awakening of the Yukon River interrupted their idyll. The Yukon was once again a great lane of traffic. As each new boat came around the last bend in the river, the dogs began to bark and the townspeople gathered along the riverbank. Steamers arrived at all hours of the day and night, each one delivering old friends, mail, or new possibilities. On board one of the first steamers to reach Rampart was future playwright Wilson Mizner, a friend of Wyatt's who was apparently exercising his creativity by running various confidence schemes that victimized miners. Josephine did not fail to notice that Mizner was standing next to an attractive female passenger, true to his reputation as a ladies' man. He and Wyatt began shouting back and forth to each other even before the boat docked. Wilson joined the Earps in their cabin, dining heartily on steak and potatoes that he swiped from the ship's kitchen ("to keep you alive until another boat comes in") and briefing them on his plans for his next stop: Nome.

Josephine had already noted the beginning of a mad migration to this northernmost outpost. People were scrambling for transportation to St. Michael, the best jumping-off point for the remote northern gold fields. Tex Rickard had already left. In fact, the Nome gold rush was kicked off in his saloon, a large circus tent in St. Michael. As Tex later told the story, a young prospector walked in and threw a heavy pouch (a "poke") of gold on the table. The usual clamor of betting and arguing hushed. Asked who he was and where he came from, he gave his name as Jafet Lindeberg and said, "I come from Nome, and there you will find more of the same stuff."

Rampart would soon become another ghost town, Josephine realized, as the previous year's exodus from Yuma played itself out again. Suddenly, an opportunity presented itself: Wyatt received an offer to run one of the Alaska Commercial Company's businesses at St. Michael, where they had stopped briefly the previous summer. It wasn't Nome, but they would be one step closer to the real action.

Josephine considered the prospect of leaving Rampart. She had enjoyed being part of an inclusive, intimate community that was fused together by the perilous winter conditions. That winter was among the happiest times of her life with Wyatt, and Rampart was the only place where they left behind close friends. So it was with mixed emotions that Josephine waved good-bye to Captain and Mrs. Mayo and their children as the riverboat steamed away, carrying the Earps and other passengers to St. Michael: "We left our little cabin by the brook, left our garden and our summer's ice. It was a sad time for both of us for we knew that there was no scenery besides ocean and tundra and barren mountains where we were going. We had learned to love the wilderness of the Yukon." But the Earps had come to Alaska to make money, not friends.

JOSEPHINE COULD BE softhearted and occasionally nostalgic. But if she had any regrets about leaving Rampart, or any other choices she made in her long life with Wyatt, she buried them deeply. The road ahead was all that mattered. The horizon was endless, always beckoning her forward. She turned her back on Rampart and focused on the next chapter.

Back on the Yukon River, Josephine felt herself to be an Alaskan veteran among a high-strung crowd of passengers who watched nervously as the boat made its way around each great bend of the river, much of it shallow and muddy. Recent storms had crushed thirty boats that year. They gasped as the expanse of the Bering Sea burst suddenly before them.

This was the crazy summer of 1898, when Klondike fever clustered around a single idea: Nome.

Wyatt's new boss, Mr. Ling, met them on the dock of St. Michael. He painted an alluring picture of the fortune to be made there, which consoled Josephine for the delay in getting to Nome. Ling directed them to a hotel owned by the Alaska Commercial Company. Josephine missed the company of their many friends and their snug little Rampart cabin but she sank gratefully back into comfort she hadn't experienced since leaving San Francisco.

St. Michael was a small, active outpost for the company, well situated for boats headed up to Nome or back "outside" to the United States. This year's crowds were larger and far more frantic than the ones Josephine had observed the last year, when she waited in St. Michael for the *Pingree* to be finished. The news about Nome had rippled throughout the entire Klondike region. Thousands of prospectors were so desperate to get to Nome that they pooled their last dollars to build small boats and row 1,800 miles across Norton Sound.

From her hotel window overlooking the port, Josephine watched the steamers and tiny rowboats crossing Norton Sound and the methodical loading and reloading of gold shipments sent down from Dawson to be shipped "outside." The lineup of people and groceries that snaked along the waterfront were signs of the ringing cash register that guaranteed Wyatt's financial success, as were the rows of beer barrels rolling toward Wyatt's "canteen." Wyatt kept 10 percent of a daily take of about $2,000. Beer was a dollar a bottle, and cigars sold for fifty cents each. The company owned his store as well as almost everything in St. Michael: the warehouses, post office, fur trading house, blacksmith shop, bathhouse, paint shop, powder house, most of the dwellings, the agents' dining room, the laundry, office building, and water tanks, and the most important and expensive structure in town: the wharf.

Despite the influx of cash, Josephine was dissatisfied. Surely Wyatt was destined for greater things than making a quick buck selling liquor. At least this retail business was a far cry from the hard-drinking, no-holds-barred consumption of Tombstone days. He was selling mostly beer, and no alcohol was actually consumed on his premises. With some prissy delicacy in her language, she noted "this custom [of no consumption on the company premises] also cut down the need for the services of an arbiter of such difficulties as used to rise in the saloons of Tombstone and Dodge City." In other words, less booze, no brawls, no bouncer.

The weather was warm, but Josephine was immediately thrust back into the deep freeze of social isolation, so repellent after Rampart. Wyatt's world of saloons and gambling and shady ladies pushed her out again to the distant margins of a disapproving society. With no homemaking responsibilities or social engagements, she filled her time with visits and walks about town, sometimes in the company of a new friend, Sarah Vawter. Mrs. Vawter and her husband Cornelius were from prominent Montana families that had fallen on hard economic times. The two women strolled along the waterfront promenade together until Cornelius was named the new U.S. marshal and the Vawters joined the throngs bound for Nome, leaving Josephine alone again.

With a flood of exuberant letters, Tex Rickard urged Josephine and Wyatt to join him in Nome. His extravagant promises befit the man who would eventually become the greatest sports promoter of his era: deriding Wyatt's steady income at St. Michael as "chickenfeed," Tex boasted that the really big money was in Nome. "Let everything go there in St. Michael," he insisted. Tex was managing Nome's busiest saloon, the Northern, which was already crowded with roulette, poker, and faro tables. That, plus the lucrative sale of alcohol by girls who danced with the male guests, made the Northern highly profitable. Tex was not worried about competition. Come to Nome and open another saloon, he urged Wyatt.

They had just arrived, but Josephine felt the frustration of being "merely on the outskirts of adventure, selling refreshments to those who were hurrying to the center of it." She was impatient to "get hold of a pan" herself. Although Tex's arguments were persuasive, she did try once more to be the practical one and encouraged Wyatt to weigh the advantages of a few more weeks of easy beer and cigar profits. Besides, everyone was trying to get to Nome, so they might not even get tickets, Josephine reasoned, half hoping, half fearing, that they would be lucky. She hedged with a suggestion that they take a short trip to Nome, just to look around—that is, if they could get a spot on a boat.

Her doubts were suddenly swept away by momentous news: gold had been found on the beaches of Nome, a barren coastline that became literally "a golden strand." The fairy tale was swiftly validated by word of mouth that spread rapidly along St. Michael's waterfront.

"Better get ready," Wyatt advised, after a quick scouting trip to the St. Michael dock. "We're leaving on the *Saidie* tonight."

JOSEPHINE ARRIVED DURING Nome's first summer as a boomtown, just a few weeks after the discoveries on the beach. Long suspected as a rich source of gold, Nome's extreme northern location, remote even by Alaskan standards, kept it off-limits. In this gray and barren landscape, there was not a tree between Josephine and the North Pole, nor another to the south for sixty miles. Until now, prospecting for gold had demanded hard labor and significant capital. Change would be radical and swift for the former fishing village and trading post. As Rex Beach noted in his 1905 novel about Nome, *The Spoilers*, "where a week before mild-eyed natives had dried their cod among the old bronze cannon, now a frenzied horde of gold-seekers paused in their rush to the new El Dorado."

Nome's place in the modern history of gold prospecting traced back to engineer and Civil War veteran David Libby, who found signs of gold while he was stringing telegraph wires across Alaska. A few months later, nearly dead with hunger and sun exposure, he was only too happy to get back to the States—especially when the rescue team informed him that during his absence, the company had abandoned the project that nearly killed him. Some thirty years later, news of nearby Klondike strikes revived his interest. With a team of engineers, he returned to his original site and triumphantly located the first major gold strike on the Seward Peninsula. Libby's announcement attracted attention from Jafet Lindeberg and his companions, the "Three Lucky Swedes" who followed Libby with major strikes near a large mountain rock that resembled an anvil. The Swedes became celebrity millionaires overnight, and their symbol briefly identified the growing settlement as Anvil City.

According to local legend, Anvil City became "Nome" when Eskimos answered "Kn-no-me"—"I don't know"—when asked for the name of the region. An alternative explanation blamed a fifty-year-old spelling mistake. In 1850 a British naval officer on a ship off the coast of Alaska noted that his map lacked a name for the prominent point, and wrote "?Name" next to the spot. Later cartographers interpreted the question mark as a "C," and thus christened the place as Cape Nome.

NEWS OF THE first Nome strikes traveled with remarkable speed: within one month, three hundred additional claims were recorded. Hardy miners who had remained in Alaska during the winter of 1898–99 staked the first claims, but as soon as the Bering Sea was navigable, thousands of miners began to arrive, some from the States and others from other parts of the Yukon. By early spring, the first boats left Dawson, with some passengers standing the entire way. Rex Beach described the eager prospectors "like a locust cloud, thousands strong, settling on the edges of the Smoky Sea, waiting the going of the ice that barred them from their Golden Fleece, from Nome the new, where men found fortune in a night."

What began as a mad scramble quickly became chaotic and dangerous.

Although the mining laws held that citizenship was irrelevant, the Swedes' claims were hotly contested on the grounds that they were not Americans. Claim jumping and lawsuits were soon rampant. Ignoring the requirement to demonstrate the presence of minerals before filing for a specific location, "pencil and hatchet" miners posted hastily written notices and carved their marks haphazardly. Large tracts of land were held speculatively.

"Lawyers are thicker [in Nome] than anything else except gamblers and sporting women," observed one resident. Litigation "has passed the stage of novelty," warned a local reporter. "It has become a matter of routine monotony that is growing more and more abhorrent to the public with every succeeding 24 hours." Powers of attorney were routinely abused, and the authorities responsible for affirming the legality of claims were accused of self-interest and double-dealing. A formal petition was sent to Washington with allegations of corruption and warnings that the mining districts were in danger of a major rebellion.

The situation grew more volatile as overcrowded boatloads of fortune hunters arrived each day, only to encounter inadequate housing and a lack of supplies. Most infuriating to the arriving miners was the discovery that "from sea-beach to sky-line the landscape was staked." After weeks of legal wrangling and rising tensions, some of the disgruntled miners banded together and announced a meeting for July 10, 1899. Their plan was to overturn the Swedes' original claim and open up the region. Hundreds of men gathered in a meeting hall, with others stationed miles away, waiting to light a bonfire on Anvil Mountain that would signal open season again on mining claims.

With not enough food, not enough work, and not enough gold, Nome was on the verge of civil war. Violence that night was avoided only by quick action of the local military leadership, which dispersed the meeting at bayonet point. Soon the miners had much more to distract them.

. . . .

WAS IT TWO soldiers digging a well? Or was it an old Idaho prospector down on his luck who was sifting sand in desperation? Other fanciful geological theories about the discovery of gold on the Nome beach attributed the miracle to an atmospheric phenomenon in mid-July 1898 that caused gold to come in with the tide like driftwood; or the fable that "a golden lake" within the Bering Sea flooded the sand each day with a new coat of precious metal. Whatever the explanation, beach miners were soon gathering enough gold to earn twenty to eighty dollars per day with minimal effort. Ordinary prospecting required heavy equipment and backbreaking labor, so this discovery seemed like magic: to be able to walk to the edge of the shore, swirl water into a metal pan and discover gold! Anybody with a pie-shaped tin could pan for gold—and everybody did. Even women and children could be alchemists. Unemployment in Nome ceased immediately.

By the end of August, some two thousand people were working on the Nome beaches.

Speculation ran wild. Was this a great find? Or a terrible and dangerous waste of time and resources? The rest of the United States heard the siren song of the golden beaches that summer and tuned out the warnings. "Truth about Cape Nome: Gold is there but All Claims are Staked" reported the *San Francisco Chronicle*, writing from "Anvil City" in July 1899. While there was plenty of gold, the writer went on to admonish the would-be prospector that "there is no chance for newcomers. By the time they could reach here from Seattle or San Francisco and get back into the country . . . it would be winter, and wintering here is almost impossible. Nobody intends to stay during the winter months if he can get out. The coast is fearfully stormy and the cold is intense." The reporter noted that slogging through the surf was dangerous and expensive, that wages were low, that there were no horses, no wood. His advice: stay away.

No one listened. That "placer gold," twinkling so tantalizingly on the earth's surface, would surely remain plentiful. It seemed that everyone in America longed to see this "poor man's paradise," where the streets were covered with mud two feet deep, but the beach was laced with treasure—and it was public property. Some effort was made to impose a tax on the beach miners, but that idea was quickly abandoned. In the absence of clear legal precedent, the miners themselves devised a solution. Meeting in Tex Rickard's saloon, the miners thrashed out a crude measure of "foot possession" to mark their spot on the beach, i.e., they could mine as much ground as a shovel could reach in a circle from the edge of the hole in which they worked.

IT TOOK A little more than one day to get from St. Michael to Nome on the creaky, overcrowded *Saidie*. From a distance, Josephine saw a shoreline shrouded by an unusual thick mist that hugged the ground like swirling snow, despite the warm weather. As the boat came closer, the fluttering waves of mist materialized into white tents staked along the fifteen miles of Nome's shore. As she drew closer, she saw people and equipment beyond the tents, but no harbor or docks extending into the water.

Nome seemed relatively accessible—at least until Josephine and Wyatt tried to get ashore. Entering Nome required a three-step maneuver that was a harbinger of the unexpected difficulties of daily life in this most unusual place. As the *Saidie* pitched from side to side, Wyatt climbed down a rope ladder and then jumped into a small boat that was pulled up alongside, and waited for Josephine, who descended a few minutes later. Rowers struggled to hold the boat steady and then navigated to shore as the tide pulled in the opposite direction. Finally, Josephine heard the welcome sound of sand grinding beneath the boat. But they were still some thirty feet from shore. As the boat bobbed up and down in the breakers, men in rubber boots waded out to them. Wyatt surged ahead through the shallow water, while Josephine and other women climbed aboard the sturdy backs of men who carried them individually to shore. The Bering Sea was now behind her, and Nome beckoned ahead, two blocks wide and five miles long.

On Front Street, the main thoroughfare, she found it nearly impossible to walk without stumbling two feet deep in mud. There were no suitable hotels yet, but Josephine found one of the few wooden shacks on "the spit" a few minutes away from the main street, slightly better than a tent.

Nome was treeless, and also sleepless. The air seemed to vibrate with the constant clamor of saws and hammers. Some two hundred wooden buildings were under construction, all built from wood imported on the boats that now arrived constantly. Josephine saw no stores, just a foul-smelling fur depot, and the warehouse of the Alaska Commercial Company. There was little in the way of municipal services: basic sanitation was almost nonexistent; sewage emptied into the river that was the only source of drinking water. Along the waterfront, tickets for public toilets were sold at ten cents each, or three for twenty-five cents; the latrines were built on pilings, which were flushed by the tide. Typhoid, bloody dysentery, and pneumonia were common. When an epidemic of smallpox broke out, a hospital was constructed in haste, with an isolation area for infected patients.

Nome was struggling to become a habitable city as the population swelled to 5,000. That summer, $1 million would be taken from the beach. Everything was being done for the first time, including burying the dead. A cemetery was hastily designated on the outer rim of the tundra, just west of the city limits. A visiting soldier described the macabre process of carving out one of the first graves: When the men finished their arduous preparation, they went to town to collect the body, only to discover upon their return that they had been "jumped." Another corpse occupied the grave, now filled in and marked by a headstone.

In spite of the general ugliness and discomfort, Josephine reveled in the excitement of Nome and without hesitation agreed that they should not return to St. Michael. Nome was bigger in every way than St. Michael, and would likely be even more profitable. For Josephine, Nome was all about the "thrill of a new gold camp," and the "adventure of the thing." At last, she would be at the center of the action.

Wyatt formed a partnership with his friend Charlie Hoxsie and built a new saloon called the Dexter. Miners needed to drink, to gamble, and to enjoy the company of women: that is how Wyatt knew how to make money, "mining the miners," as he said to his brothers. Every minute of that summer was devoted to getting the saloon up and running. This would be Nome's first two-story building and its biggest structure to date, named after John Dexter, whose trading post had been the central gathering point for the pioneers of the Nome gold rush, in the earliest days of the Three Lucky Swedes. Wyatt and Hoxsie purchased "a bar-room license to engage in the sale of intoxicating liquors at retail." Aided by Wyatt's celebrity, almost as potent in the northern frontier as it had been in the American territories, the Dexter was an immediate success and enjoyed what the newspapers called "a liberal patronage" during the years of Wyatt's management, in friendly competition with Tex Rickard's Northern Saloon as the most popular spot in Nome.

Josephine was torn between her vague disapproval of Wyatt's saloon keeping and her frank appreciation for its "lavish financial returns." In St. Michael, she had comforted herself with the dubious distinction between selling alcohol and encouraging its consumption. In Nome, she could have no such illusions. She rationalized anew: the Dexter, this "better class" saloon, served an "important civic purpose" as the local clubhouse, the town hall, and the forum, where men could arrange political campaigns, transact business, and enjoy social contacts. Into the doors of the Dexter walked any important resident or guest, from writers like Rex Beach and Jack London to mining engineer and future president Herbert Hoover to future prizefighter Jack Dempsey. Wilson Mizner was there too, often entertaining a crowd at Considine's Hall, singing a lusty version of "Ben Bolt." "We met and hung out in saloons," Wyatt later observed with a smile. "There weren't any YMCAs."

The Earps were getting rich, but the Dexter's popularity did nothing for Josephine. In fact, Nome would later pass an ordinance forbidding the presence of women in saloons. She could have entertained herself in other ways: women were working their own claims, and had seized the right to vote in the city's elections for its first mayor, city council, and chief of police. However, Josephine was never interested in politics. What passed for high society in Nome was already closed to her. As the wife of Nome's most famous saloonkeeper, she was invisible, caught once again in an unpleasantly familiar netherworld.

Nome matured with an intensity born of nearly twenty-four-hour days and the superheated atmosphere of gold. However, in the more sober corners of the community, winter was a subject of growing concern. Nome was getting a little chillier each day. Most people were still living in tents and shacks, facing a severe shortage of fuel, food, and medical supplies. Newspaper editorials warned residents to prepare for the winter—or leave: "This is the time when those who intend to winter in Nome should begin skirmishing around for eight months' supplies. . . . Nome cannot afford to be burdened with indigents this winter, and any one who does not see clearly whence his or her support is to come from had best get out before it is too late. We can hardly imagine a more uncongenial place to be broke in than Nome for the eight months commencing early in November."

As the temperature plunged, panic hit those who had not yet struck it rich. For every successful prospector, there were at least three who had failed. Some 1,000 miners were broke and homeless in Nome, where everything cost up to five times as much as it did back home. Some vowed to stay, but most of them took advantage of government subsidies to board a steamer back to the States.

Charlie Hoxsie advised Josephine and Wyatt to leave their investments, including the Dexter, in his hands for the winter. Josephine was wary of the increasingly strong winds and driving rains that foreshadowed the approach of the great ice pack marching down from the north, a "wall of ice shimmering like diamonds," as high as sixty feet, and advancing as rapidly as three miles an hour. Still more horrifying to her imagination was the fearful sound of the ice pack bearing down on Nome, a head-filling groan that Charlie and other veterans recollected with a shudder. "You can't hear yourself think," Charlie warned, as the icebergs crashed against each other, creating a maddening din that ended suddenly in an ominous moment when the sea froze and empty silence settled over the landscape.

Despite Charlie's dramatic predictions, Josephine wanted to remain in Nome, hoping perhaps for a reprise of their idyllic winter in Rampart, but Wyatt overruled her. He had been unsuccessful in finding a suitable home for them, "no nice warm log houses" as they had in Rampart. Between his concern and the fears on everyone's face, Josephine capitulated. They decided to spend the winter in Seattle and San Francisco, where they would outfit the Dexter more grandly and return to Nome as soon as the Bering Sea became navigable.

They were almost too late to board the *Cleveland*, the last boat to leave Nome. Wyatt bribed two men to relinquish their small stateroom, just big enough for the two of them and a collection of furs and fine Irish woolen shawls, robes, and blankets purchased by Josephine. The trip was a nightmare. First came the discovery that their bodies and clothing, plus everything in their stateroom, were crawling with lice. Her beautiful furs and woolens were thrown overboard. Then they encountered a storm so terrible that Josephine begged to get off the boat, though they were in the middle of the Bering Sea. As they itched unmercifully, the boat rolled and pitched amid mountain-high waves beneath a "black howling sky." When they finally reached Seattle, nine days late and already given up for lost, they had only the clothes on their backs and a few things that they hoped would survive fumigation, such as Josephine's treasured lynx cape. Still, they fared better than the passengers of another boat that left at the same time; the *Hera* took twenty-eight days to make the week-long trip, during which "two men died from starvation and others were half crazed from want of food and water . . . the majority of the men were so weak that they could not carry their gold dust ashore without assistance."

Despite the horrors of their voyage, the passengers of the *Hera* were united in "declaring Nome to be the greatest camp on earth." Many of them were already planning to return in the spring.

AS JOSEPHINE STEPPED onto dry land, grateful beyond words to be off the boat, she encountered a city that had a single-minded focus. Seattle was Nome-crazy. The newly formed "Cape Nome Information and Supply Bureau" bombarded pedestrians with ads promoting Nome underwear, Nome tents, Nome medicine, even special hats like the Reed's Blizzard Defier Face Protector, which promised that "whether your nose is long or short, wide or narrow, inclined to be roman or retroussé . . . the wearer can see, hear, breathe, talk, smoke, swear, chew, or expectorate just as well with it on as off."

One winter in Rampart had been enough to convince Josephine's neighbor Erasmus Brainerd, now back in Seattle, that Alaska was the new frontier. Seattle had missed one gold rush, Brainerd argued, and might never have another one. The city was losing ground to Tacoma and Portland—and all three cities were in the giant shadow of San Francisco. The national picture was still worse, a sickening combination of unemployment, business failures, and bank closures.

Alaska was a gift of the new millennium. Seattle successfully petitioned the federal government to establish an assay office where miners could have their gold tested and valued, cutting out the longer trip to San Francisco. The city's traditional businesses also prospered, as it became a primary source of lumber for the treeless north. Construction business boomed as profits were invested in the urban infrastructure.

The Alaska gold rush, especially the explosive summers of 1899 and 1900, changed the history of Seattle. Brainerd and his colleagues launched a national media campaign to promote Seattle as the gateway to the fabled riches of the Alaskan goldfields. It worked: the city drew the majority of

Nome gold seekers to its stores and wharves. The amount of money passing through its banks tripled. In three years, Seattle's population increased by about 30 percent.

Josephine and Wyatt split the winter months between Seattle and San Francisco. The local newspapers took note of their return: the *San Francisco Examiner* reported that Wyatt was "making money perhaps faster than he ever made it before" and predicted that "if business runs with him next summer as it did after his arrival in the camp, he will be able to retire with all the money he desires." Readers of Seattle's newspapers read about Wyatt as the "celebrated sheriff from Arizona" and a "quiet sort of individual, good natured, does not talk much," though some reporters condemned him as "a bad man" and eagerly reprised the Sharkey-Fitzsimmons debacle as evidence of his corruption. Back in Tombstone, the *Epitaph* noted that their most famous former resident was back from Nome and would soon be opening a combination club and saloon.

Hoping to trade again on his celebrity, Wyatt opened a new gambling and prizefighting club in Seattle's Tenderloin district, but after a flurry of publicity and eager crowds, he ran into a brick wall of uncooperative police and politicians. Seattle was not Nome, and stirred up by prohibition enthusiasts and competitive saloonkeepers, the city was increasingly hostile to Wyatt. The state of Washington initiated legal proceedings against him for gambling and selling alcohol, and the club's furnishings were confiscated and torched. The city's temperance craze passed quickly, however, and soon everyone was back in business—except Wyatt. He and Josephine spent the rest of that winter buying glamorous furnishings for the Dexter—expensive thick carpets, mirrors, carved sideboards, and draperies.

Back in Nome, the winter passed slowly. The city's residents had only the sketchiest of information about what was going on in the rest of the world. Newspapers could ask but not answer the question, "Is the Philippine war still on or has it at last been settled?"

The people of Nome seized any opportunity for winter celebration. "First child of Pure Caucasian Blood Born in Nome" announced the *Nome Nugget* on January 6, congratulating Mrs. Ginivin on the birth of the city's first white child, a ten-pound boy who happened to be born on New Year's Day. (Unfortunately, Mr. Ginivin was still on the "outside," having left in the fall.) Other entertainment came from visitors such as Rex Beach, who performed in a minstrel show in which "he was the chief burnt-cork artist, furnishing the audience with more merriment than ordinarily falls to the lot of the Nome citizen during his period of winter hibernation." Nome residents loved to bet on anything, from the date the ice broke up to the outcome of elections and dog races. They even liked to bet on betting—as they did when they ran a contest to decide the most popular faro dealer in town.

With regular business in suspended animation, Josephine's friends initiated civic projects during the long winter freeze. For Christmas Day, Tex Rickard fed the poor with seven hundred turkey dinners, establishing an annual tradition during the years that he lived in Nome. Beloved for his charitable ways, Rickard was asked to serve as Nome's first mayor. Charles Hoxsie was similarly philanthropic and was considered "one of the best known and most popular men in Nome . . . an enterprising citizen, a whole souled and altogether good fellow."

By March, Nome residents began to talk feverishly about rejoining the outside world. A few hardy souls arrived from other parts of Alaska by dog teams, skates, even bicycles, but the waterways were still blocked by ice. The *Nome Nugget* offered a $10 prize to the person who guessed the day and hour of the arrival of the first steamer. Estimates of some 100,000 new residents filled the newspapers. At least some of these new arrivals were anticipated with fear: the Northwest Mounted Police warned the local officials that "the most criminals ever known on this continent" were headed to Nome from all over Alaska. The attorney general's special agent predicted that the summer would bring Nome "the worst aggregation of criminals and unprincipled men and women that were ever drawn together in this country."

While all of Nome was eagerly looking out to sea for the first boats, with anticipation and some anxiety, a much bigger mass of people was gathering at Seattle, like an army preparing for an assault. One of the most inaccessible places in the world during the frozen months, Nome was now advertised as an exotic summer destination, suitable for tourists as well as treasure seekers, with gambling to while away the hours on board the comfortable steamers. The trip cost about $75, which included 1,000 pounds of freight for the entrepreneurs on board. The *San Francisco Chronicle* predicted excitedly that Nome would soon be "A Modern City," with electric lights, power, streets, railways, and telephone lines. Business owners invested in steamers to connect Portland and Cape Nome. But as Josephine was soon to be reminded, nature was not easily tamed. The Bering Sea was still clogged with unmapped blockades of frozen water. Boats were often forced hundreds of miles out of their way in search of an open channel, and could be tossed around like toys by the unpredictable movement of the ice. The so-called holiday cruise could take as little as a week or as long as two months.

Huge crowds filled the Seattle waterfront to see the first travelers depart in May 1900. On May 20 alone, four ships left Seattle, filled to capacity with as many as seven hundred people and loaded with thousands of tons of mining machinery and general merchandise. Other boats carried dismantled theaters, gambling halls, saloons, hotels, and restaurants—everything needed to construct an "instant civilization."

Josephine and Wyatt left from Seattle on the *Alliance*, with the luxurious accessories for the Dexter in the hold. By now, the rest of the world knew what they knew: that Nome was destined to become "the greatest mining camp city the world has ever known."

The time passed pleasantly enough during the early days of the trip. Josephine entertained herself with gambling, losing enough to annoy Wyatt. The Bering Sea became impassable just when they reached the port of Unalaska. The harbor town was crowded with people headed for Nome; among them, Josephine discovered many friends from her previous trips to Alaska. The biggest surprise came when Josephine heard a familiar voice calling "Aunt Josie!" There was her niece Alice with her husband Isidore, on their way from Oakland to Nome, where they had hoped to surprise Josephine and Wyatt. Isidore was bringing samples of clothing to sell in Alaska. An attentive and affectionate aunt, Josephine made plans for their time together in Nome.

They were soon back on board the *Alliance*, but not out of danger. As the ship skirted a menacing cluster of icebergs, Josephine noticed an alarming sound as the engines strained to negotiate lethal ice floes clustering around them like large, threatening animals. She awoke one morning to an eerie stillness. The engines had stopped completely. In place of the gray water of the Bering Sea was an impenetrable ice field as far as she could see, "a glistening white sheet, with spires and towers, hummocks and peaks." The ship was clutched by the ice for another full week, as rations dwindled and tempers frayed. Finally distant boats began to stir, and the *Alliance* penetrated an open lane.

She was again within sight of the Nome coastline, close enough to take stock of dramatic changes. Everything was on a far bigger, noisier scale: where there had been a few boats in the water, there were now scores of ships lying off the shore, plus some wrecked hulks left over from winter storms. Instead of the little dories that had pulled alongside her boat last year, a motley flotilla of barges, tugboats, rafts, and rowboats hustled back and forth, loading and emptying enormous loads of cargo. Hundreds of cattle were being pushed into the water and herded to shore in a watery roundup. An army of Paul Bunyans plunged into the water to offer broad backs to Josephine and the other women climbing down ladders. Wyatt stayed behind to supervise the transfer of their precious cargo, while Josephine clung to the human ferry that deposited her on the beach. The air was filled with riotous sounds as each new boat blasted a shrill whistle to announce its arrival and every ship in the harbor answered in return, the passengers cheering and applauding, many having been stuck on ice for weeks.

For one brief summer, Nome became one of the busiest and oddest seaports in the world, where the last mile of freight delivery cost almost as much as it did to traverse the two thousand miles from Seattle.

Once on dry land, Josephine faced a scene of unimaginable chaos. The beach was barely visible beneath thousands of tents that almost touched each other, leaving the narrowest of passageways between them. Small mountains of worldly goods broke the line of tents, each pile challenging its owner to carry it away faster than a thief or a storm. Hundreds of dogs raced furiously about. Baggage and freight were piled high on the beach for a distance of several miles: a jumble of pianos, coal, narrow-gauge railway tracks, lumber, tents, stacks of hay, bar fixtures, washtubs, roulette wheels, stoves, liquor, sewing machines, and mining apparatus. Despite the density of people and belongings, wagons lumbered along the beach, hawking baked goods, clothing, and mining supplies. Men with go-carts were hauling loads; others were carrying trunks on stretchers or on their backs. Water was delivered from a wagon bearing a large water tank, five gallons for twenty-five cents. Crazy gold-panning contraptions were in constant motion.

For years after arriving in Nome, people looked for worldly possessions that were lost on their first day. Sometimes personal baggage was unloaded from the boats immediately; sometimes it was kept on board and sent later. Nor did businesses awaiting deliveries fare any better than hapless individuals. One grocer was expecting $10,000 worth of canned goods when an unexpected dunk in the Bering Sea washed off all the labels; it was sold as "mystery food" for ten cents a can.

Most people arrived without a plan other than a fierce determination to grow rich on Nome's golden sands. "Imagine a heterogeneous mob of 23,000 people landed on a beach," said one visiting engineer from London, "huddled together on a strip of beach 60 feet wide . . . without any sanitary arrangements; sleeping on almost frozen ground at night, broiling in a hot sun by day; and you have Nome as it was." The first night was a particularly miserable rite of passage. Lucky newcomers who had the foresight to bring tents scrambled to find places to pitch them. Those without tents waited uncomfortably for the ships to unload their possessions, sleeping on the sand in the twenty-four-hour daylight with no shelter and relief from the crowds and the constant clang of construction.

Josephine was dazed, but Wyatt never took his eyes off their belongings until everything was stacked safely onshore. Steamship companies claimed that their responsibility for delivery ended when the goods left the ship, and Wyatt had invested too much money to be careless about the expensive furnishings he had purchased for the Dexter.

Josephine was used to life in a boomtown, but Nome's northern latitude generated an intensity beyond anything she had experienced. There were thirty days of ceaseless activity that summer, with as many people on the streets in the middle of the night as there were at lunchtime. Stately business blocks sprang up out of nowhere, and rich and costly interiors were installed and ready for customers in a few hours. There were fancy hotels, beer gardens, about a hundred saloons and gambling houses, and several newspapers. An empty lot one day would be reclaimed from the tundra and filled with a three-story structure the next day, its storerooms filled with bright new goods. Within weeks, Nome grew to a city of over 20,000.

Front Street was still unpaved and filled with people rushing about, elbowing each other for room. The dense black mud was so deep that when a wagon approached, everyone ran for the closest doorway to avoid being splattered. Many of the buildings were identified by wooden signs with a single word, SALOON, and had a dance hall in the rear. It was not unusual for someone walking along the street to be pushed by the crowd into a saloon, and it was often easier to get in than out.

Public health conditions had only slightly improved from Josephine's first summer. Several ships in the harbor flew the yellow flag of smallpox quarantine that kept all passengers on board. Public water closets were no match for the demand of the wildly growing population. Garbage was collected weekly; in winter, refuse was carried on the ice to the three-mile limit, where the summer currents carried it to the Bering Straits and then out to the Arctic Ocean. The sanitation business was so profitable that the garbage collector's daughter was reported to be the best-dressed girl in Nome's first school, outfitted with a beautiful ermine coat shipped from New York.

But the beach at Nome, once "thick with gold," had reverted to prosaic sand. One year after the momentous discovery, the pans in the hands of feverish amateurs were yielding flecks, not nuggets, which were hardly worth the effort to retrieve. Unemployment was growing, and emotions ran high, veering between disappointment and hope. "This is a hot town and going to be hotter," wrote one prospector trying to sustain his optimism. "Hundreds of men sitting around not knowing what to do and hundreds more coming in every day. Labor is a dollar an hour, but one has got to be a husky or strong man to get a job." In one night, he reported, three men committed suicide and two men shot each other on the street.

As Josephine had observed in Tombstone, business boomed even after mining opportunities peaked. So did crime: Nome was approaching the point where predators outnumbered prospectors. The same newspapers that wrote breathlessly about Nome in 1899 were now warning about a "reign of terror" that would leave Nome with little or no gold being taken out, overrun by gamblers and disreputable characters from other parts of Alaska and all over the United States. Guns were commonplace, martial law was imposed, and Nome's highest-ranking military official wrote to Washington that there was "no effective civil organization for protection of life and property." The owner of the Alaska Commercial Company visited Nome several times that summer, but the company's internal memos warned that the city was "filled with the riffraff of the country, law and order were disregarded and honest men could scarcely make a living." One specialty crime involved cutting a hole in a tent and pumping in a powerful dose of chloroform before robbing the sleeping occupants.

Despite forbidding headlines back in the States—"Murders Frequent and Gold Very Scarce"—people kept coming.

"The major business in Nome in 1900 was not mining, but gambling and the saloon trade," reported the *Seattle Intelligencer.* "Downtown Nome was lined with nearly 100 saloons and gambling houses, with an occasional restaurant sandwiched in between." The Dexter was at the top of the list. In its second year of operation, the Dexter consolidated its position as Nome's preeminent saloon for liquor and gambling, aided by Wyatt's celebrity and his winter shopping expedition, which created a first-class saloon fit for any city in the United States. He and Hoxsie renewed their alcohol license, which required them to swear that the Dexter was more than four hundred feet distant from any schoolhouse or place of religious worship. In the Nome of 1900, that was not hard to do.

Josephine suppressed her objections about Wyatt's occupation as gambler and purveyor of alcohol. They were making too much money. Former patrons of the Dexter would recall sky's-the-limit poker games, with as much as $500,000 in gold riding on a single hand. Nor did Josephine express misgivings about prizefighting, which had become a central feature of Nome's sporting life and a great moneymaker for Wyatt and Tex Rickard. Although it had already been declared illegal in some states, with well-publicized arrests in New York, Cincinnati, Denver, and Kansas, prizefighting was still hugely popular in Alaska. Nome became a regular stop on the circuit, and the local newspapers carried blow-by-blow coverage accompanied by pen-and-ink illustrations. Wyatt often refereed the fights, and one can only imagine Josephine's reaction when she heard that Bob Fitzsimmons would be fighting in Nome in advance of another bout with Sharkey.

Nome's newspapers faithfully chronicled events attended by the "Nome

400," which included the high-ranking military officers, business owners, and those who would be considered high society on the outside. Rex Beach mocked this group in *The Spoilers* when his heroine Helen rejected what passed for Alaska society: "They talk scandal all the time. One would think that a great, clean, fresh, vigorous country like this would broaden the women as it broadens the men, but it doesn't." High-stepping around the mud, the saloons, and the smallpox, the Nome 400 found time and place for fancy parties and formal balls at the Nome Standard or the new Golden Gate Hotel. This was a world to which Josephine was never invited, a social whirl with deckle-edge dance cards inscribed by partners for the waltz, two-step, minuet, or quadrille, followed by lavish dinners with oysters and fricassee of lamb and Mumm's champagne.

Josephine was barred from Nome high society, but her friend from St. Michael, Mrs. Vawter, often attended these events with her husband, Cornelius Vawter. Their friendship was strained when Marshal Vawter and Wyatt found themselves in opposition. Remarkably, just as Johnny Behan had once passed over Wyatt for the deputy sheriff's job in Tombstone, Cornelius Vawter refused to appoint Wyatt as deputy marshal of Nome.

The narrow and stratified society of Nome stood in stark contrast to the cosmopolitan versatility of San Diego, San Francisco, or Seattle, or the welcoming community of tiny Rampart. Josephine sought consolation in the company of her family and old friends. In addition to her niece and nephew, her brother Nathan had arrived from San Francisco, and Wilson Mizner and Rex Beach were lively companions. Sidney Grauman was there as well; from his humble beginnings as a newspaper salesman, he was on his way to becoming an entertainment mogul and had invested in the local theater and opera house.

Josephine found no more kinship in the small but influential Jewish community in Nome than she had in Tombstone. The gold rush pioneers who first initiated Alaska's Jewish services in Dawson City in 1898 were now successful merchants tied closely to the Alaska Commercial Company.

To usher in the Jewish New Year of 5661, the *Nome Gold Digger* devoted its front page to the opening of the first Nome synagogue, with about sixty celebrants, led by the former head of a Spokane synagogue and using a Torah from an Amsterdam congregation. The services would be traditional, noted the *Nome Chronicle* on September 18, 1900, because so far from home, "even reformed Jews like to revert again to the strictly orthodox faith." Claiming the distinction of being the most northern and western of Jewish communities, the so-called Frozen Chosen continued to meet weekly for Friday-night services.

Despite the general respect afforded to the Jews of Nome, the local newspapers were not above some unsavory jokes, like the prominent cartoon of a "Hebrew Bunkoed on a Whiskey deal," a hook-nosed fellow who was tricked into paying fifty dollars for a barrel filled with salt water. "Captain vot you tink!" the man complained. "Ven I hopened dot barrel, vot do you tink vas in it? Nothink but salt vater, captain! Vot shall I do, captain? Tell me, vot I shall do!"

Although the Dexter was prospering, the summer of 1900 generated serious marital tensions. Josephine may have overcome her squeamishness about saloon keeping and prizefighting, but she drew the line at prostitution, as openly tolerated and cheerfully regulated in Nome as it had been in Tombstone and San Diego.

Nome's red-light district, "the Stockade," was in the filthiest part of town. Women paid a monthly fine of ten dollars to the chief of police, who split the fee with the municipal judge and the city treasury to support services such as fire protection and welfare for the destitute. Many of the women had families they were supporting back home, or were wives trying to make enough money to get back to the States. Wearing dressing gowns, the women of the Stockade stood in a row of small frame houses with their names on the door, calling out to the men and staring at the women with smiling indifference.

Prostitution was not limited to the Stockade. Most of the Nome saloons turned into dance halls at night, with discreet bedrooms upstairs for special patrons. When the Dexter opened its own second-story "club rooms," Josephine protested loudly and publicly that these rooms were for "games of chance, not frolic." The presence of her niece and nephew undoubtedly made her more sensitive, since they would bring back to San Francisco the stories of what they had seen and heard in Nome. As she feared, her niece Alice would long remember Josephine's fury at "the whorehouse above the saloon" and how she ordered Wyatt to get rid of the women.

It was a summer that revealed more of the famous couple's weaknesses. Josephine confessed to her friend and biographer Mabel Cason that Wyatt had affairs during their time in Nome. She was alone most of the time, or with her family. She was gambling heavily—and losing. Instead of betting on the horses—impossible in Nome—she was indulging her fondness for card games, and she had trouble meeting her debts.

For his part, Wyatt had several run-ins with the law. "Some little excitement was occasioned on Front Street this morning by the occurrence of a drunken row," reported the *Nome Daily News*. Wyatt Earp and Josephine's brother Nathan were arrested in a brawl that started in front of the Dexter, when two drunken patrons got into a fight. Wyatt and Nathan tried to intervene and were taken into custody. Wyatt claimed that his actions had been misconstrued and that he was trying to assist, not hinder, the actions of the deputy marshal. Despite Wyatt's initial protests of innocence, he entered a plea of guilty and paid his fines.

As if to keep pace with their local woes, the specter of Tombstone suddenly returned with a vengeance. Wyatt's youngest brother Warren Earp was shot in Wilcox, Arizona, and the newspapers were quick to link his violent death to Tombstone. Warren's killing was an unwelcome reminder that even here, Josephine could never escape the notoriety of the O.K. Corral and the enduring interest in the Earps and the cowboys. It was no consolation that this time it was not Mattie Blaylock who was haunting her, but the events of October 26, 1881.

Inaccuracies flew. The *New York Tribune* carried a front-page story confusing Wyatt with Warren, Nome with Denver: "Wyatt Earp Shot at Nome; The Arizona 'Bad Man' Not Quick Enough with His Gun." The *Tribune* reiterated Earp's reputation as a "dead shot" and a "bully." The *Seattle Post-Intelligencer* and other newspapers reported the death of Virgil, not Warren, and repeated other old canards. Some accounts placed Wyatt and Josephine in Phoenix soon after the killing and had Wyatt embarking on a second vendetta to avenge Warren. A Chicago newspaper published a particularly ludicrous article about an Englishman terrorized by Wyatt Earp and forced to eat hot tamales in Alaska at gunpoint.

The *Arctic Weekly Sun* commented that Wyatt, unlike his brothers, "seems inclined to break the record and die a natural death." As if testing that assumption, Wyatt did manage to become embroiled in a second fight that summer, another barroom brawl with Nathan that began with a military policeman trying to stop a fight at the Dexter. "While the soldier is doing his duty, he is assaulted and beaten by Wyatt Earp and Nathan Marcus," reported the *Nome Daily News* on September 12, 1900.

Once again, Josephine found herself struggling to defend Wyatt. From the remoteness of Nome, she was pulled back to the violence of the Arizona Territory.

Tombstone brought more welcome associations to Nome with the arrival of old friends, beginning with John Clum. The former editor of the *Tombstone Epitaph* had distinguished himself as a high-ranking official in the U.S. Post Office, now posted to the Alaskan frontier. He was among the thousands of volunteers who rescued survivors and recovered bodies from a deadly avalanche at the Chilkoot Pass. Clum quickly revolutionized mail service in Nome. Before his arrival, people were standing in line for mail for up to two days; the going rate was one dollar an hour for someone to hold a place in the queue. Clum commandeered a small building and hired an army of clerks who worked around the clock to eliminate the large backlog of mail from the Seattle post office. In the summer of 1900, Nome was the largest general delivery post office in the United States.

George Parsons arrived next. His first glimpse of Nome's shoreline of crowded tents and shanties reminded him of "Old Tombstone." Parsons had come to Alaska as the representative of a mining syndicate, but he soon came to consider that summer the "worst tramp of my life" as he fought through the rain and mud of Nome. Always an admirer of Wyatt's, Parsons visited the Dexter often, which he called "the biggest drinking and gambling place here." To his pleasure, Wyatt Earp seemed undiminished, still "straight and fearless." He made no comment about Josephine.

Clum remarked on the coincidence of their three-way reunion occurring at the same time that the Nome newspapers carried the story of "Apache Geronimo Insane," caricaturing the former warrior's wild antics at Fort Sills, Oklahoma. Nineteen years before, the three friends from Tombstone had joined a scouting party that was on the trail of Geronimo. Now they were together in Nome, having "a regular old Arizona time." Clum recounted later, "It seemed proper that we should fittingly celebrate this reunion of scarless veterans on that remote, bleak, and inhospitable shore—and we did." They reminisced with pleasure, and Parsons recorded one particular evening on August 30, near the end of that memorable summer of 1900: "We had such a séance last night. That evening with Wyatt Earp would have been worth $1,000 or more to the newspapers."

Lucky Baldwin showed up just in time to play a minor role in one of the greatest frauds in American legal history. He had come to Nome to start over again. Once he had been Wyatt's wildly successful business associate and friend, and Josephine's not-so-secret admirer. He had covered her gambling debts in San Diego and enjoyed friendly racetrack competitions with Wyatt. But now he was the one who needed his old friends. Many of his ventures had failed, his real estate was heavily mortgaged, and his beloved Baldwin Hotel had burned to the ground.

Still walking about with his customary wad of $100 bills, Lucky planned to establish a mining operation and to open up a saloon with gambling equipment that he had brought from San Francisco. Wyatt loaned him temporary space, as he had for other old friends. But luck finally seemed to have deserted Lucky Baldwin. "I came up here, expecting not only to go into business largely, but to do some mining on the beach," he told the *Nome Gold Digger.* "The beach, however, is not what we supposed it was on the outside." He came too late to Nome for mining, and he was also too late for gambling: in a town with limited real estate, all the good locations were taken. Then his property was seized by corrupt local officials and held on a trumped-up charge of tax evasion. Marshal Vawter told Lucky that there was a legal claim for $2,500 against him, but that it would be released for $10,000. Baldwin turned to Wyatt for help, and Wyatt quickly arranged for a bond. But Vawter then increased his demands to $20,000—in gold dust. Loyal to his friend, and indignant at the marshal's extortion tactics, Wyatt immediately raised the gold and even supplied someone to deliver it.

Lucky was just one victim caught in the grip of white-collar crime that held all of Nome hostage during the summer of 1900. The drama began with the arrival of Judge Arthur Noyes, a political appointee of Alexander McKenzie, a well-placed Republican, who was filling all of the important offices in Nome with his cronies. In addition to Judge Noyes, the most influential of McKenzie's henchmen was U.S. marshal Cornelius Vawter.

Their grip on Nome was brief, but the three months of their reign unfolded in Nome-time, under the intense glare of twenty-four hours of daylight. First, Noyes and McKenzie manipulated the law to ambush the most valuable mining claims throughout Nome. In response, the miners sent legal papers secretly to San Francisco by steamship. Speed was essential; the cold season would be soon upon them.

In the meantime, Nome was overwhelmed by another catastrophe.

The early summer of 1900 had been unusually warm and dry. The Arctic tundra was bright with sunshine and dotted with colorful wildflowers. The winds shifted, the skies darkened, and on September 11, the rains began.

Waves invaded Front Street and pushed waterfront cabins aside like bulldozers. George Parsons described the fearful sight of "buildings and homes swept into the ocean in one tangled mess." Streets became turbulent, debris-strewn rivers. About a hundred people were killed. With winds reaching seventy-five miles per hour, many vessels were unable to ride out the storm and were destroyed. The *Skookum*, a large barge that had been constructed in Seattle and towed to Nome, dragged its anchor toward the beach, with thirteen men and tons of cargo still on board. Horrified and helpless, most of Nome watched the *Skookum* "grind her vitals on the sands of the beach." After an hour of pounding, shortly before midnight, "when the storm and surf were at their height, the *Skookum*, with a mighty crash, broke in the water." The men were lost at sea. Beachcombers salvaged what they could, especially the wood that would be so essential for the following winter's fuel.

The storm brought all of Nome together. In a moment of solidarity, in the midst of legal and political battles that made enemies of their husbands, Josephine and Mrs. Vawter found common cause. Under the headline "Help for the Needy," the *Nome News* reported, "Mrs. Vawter, Mrs. Wyatt Earp, and Mrs. Lowenstein have collected $80, which they will distribute to those who suffered in the storm. Those who are in need of help will get it if they apply to Mrs. Earp at her residence on the spit, next to the office of Dr. Tiedeman. Only women and those with families dependent on them need apply." (Nothing else is known of Mrs. Lowenstein.)

The distraction was brief. Bodies were still washing up on shore when court papers arrived from San Francisco confirming that Judge Noyes and McKenzie had grossly abused their power. But even the dramatic, hand-delivered court order had no effect. McKenzie simply ignored it, and Noyes refused to enforce it. A second messenger was dispatched to San Francisco. This time, he returned with a warrant for McKenzie's arrest.

The same political connections that installed Noyes and McKenzie as dictators in Nome protected them now. Judge Noyes was fined $1,000 and reassigned to another court. McKenzie, who knew every Republican president from Grant to Harding, served three months of a one-year sentence before President William McKinley pardoned him on grounds of severe health problems. McKenzie made a miraculous recovery, however, and lived another twenty years. Marshal Vawter's role was somewhat mitigated, since he had made some effort to protect the miners. Nevertheless, he and Mrs. Vawter were banished to the outpost of Unga, which the *Valdez News* described as "a little God-forsaken hole out on the Aleutian peninsula."

The summer of 1900 culminated in two storms—one natural, one man-made—that nearly destroyed Nome. The whitewashed skies of late September seemed to signal the end of the saga that had drawn Josephine and Wyatt and so many others to Nome. The Earps had made some $80,000 (more than a million dollars today); most of this came from the Dexter. Even Lucky Baldwin managed to recoup his investment. He offered to repay Wyatt and Josephine, but Wyatt refused, saying that he was "more than satisfied to put a crimp in the grafting of that crowd of crooks." In fact, Wyatt was one of the few who stood up against one of the worst legal frauds in American history. As George Parsons would note later about Wyatt's actions in Nome, "if anybody was undeservedly ill-treated and especially an old Tombstoner, he would find a champion in the same Wyatt Earp, who is older now but none the less gritty."

Nome was no longer the wide-open frontier. The mining town had grown as if by magic and for the briefest, most intense of summers, it justified the boast of the *Nome Nugget* to be the greatest mining camp ever known. But now the beach was worse than barren: it was strewn with the aftermath of a catastrophic storm, the last straw for many unsuccessful miners. Nome accounted for almost half of the total gold output of the Seward

Peninsula in 1900. The mining industry would continue to thrive at similar levels for the next five years, with some major discoveries still to come in 1904 and 1905. But the majority of the people who came to Nome in those early years left poorer, sadder, and probably no wiser. For some, it was a lark: "My family stayed up in Alaska about a year, then they came back," Josephine's grandniece later recalled. "They didn't get rich. I have a gold nugget that my father dug out of the mines, that I wear as a clip."

After the storm, the ticket offices were jammed with people who wanted to leave on the next boat. Many of the unsuccessful miners were penniless or in debt or trying to barter or sell the already decrepit mining equipment that was their only capital. Hundreds went home dead broke, sent back at taxpayer expense like wartime refugees. Brigadier General George M. Randall, commander of the department of Alaska, cried shame on his own government for aiding that "venturesome class of white men" while offering no support to the Alaskan natives, who lost their livelihoods and communities when their natural resources were raided by hordes of fur trappers and prospectors. "It is either a beach of gold two hundred miles long, the greatest placer gold field in the world, or the greatest disappointment of modern times," lectured the *New York Times*. "The lesson for these sad observers is that learned by the thousands who have returned home as guests of the United States—that there is more wealth to be gained by downright every-day plugging at some matter-of-fact business than can be picked up on golden beaches."

The stampede was on again, only now the direction had reversed. Everyone was trying to get home.

JOSEPHINE AND WYATT wound up their affairs rapidly. Although Wyatt's name would continue to appear on advertisements for the Dexter, he sold his ownership to Hoxsie. The mining claims that Wyatt had filed in Josephine's name were legally transferred to her brother Nathan.

They would not be there to see Nome's next phase, though Josephine might have echoed Rex Beach's flinty hero in *The Spoilers,* who scoffed, "They're puttin' a pavement on Front Street." By the next summer, Nome had been legally incorporated and enjoyed the benefits of sidewalks, electricity, a local brewery, and a new Catholic church. Gold was yesterday's story, as editorials in the States began to focus on the promising coal deposits in Alaska. Elections were held for the first municipal officers, and among the first councilmen were Tex Rickard and Charlie Hoxsie. They had become pillars of the community in a way that Wyatt and Josephine never would.

Josephine had withstood the rigors of one frozen winter, and she had survived the pressures of three stormy years. Like a nomadic fraternity, the Earps' friends were already regrouping in Nevada and California deserts. There would be other gold fields, and other adventures, but no place like Nome.

Once again, the frictions of their marriage were resolved, and they were ready to move on, together. As they left Nome aboard the SS *Roanoke* in the fall of 1901, Wyatt was reported to have said, "She's been a good old burg. Mighty good to us."

## 4 | WAITING FOR WYATT

[WYATT EARP] has just returned from Nome, where he has mining properties sufficient to make him financially comfortable for the rest of his life," reported the *Los Angeles Express*. "Mr. and Mrs. Earp will continue their journey south tomorrow and will return to Nome the coming season."

The *Express* had it mostly wrong. The profits made in Nome were not from gold mines. Nor would the Earps ever return to Alaska.

But Mr. and Mrs. Earp had at last made their fortune, enough to follow their inclinations. They loved the desert; they loved prospecting; the rootless life of temporary residences and changing landscapes suited them. Wyatt suggested the purchase of a cattle ranch, perhaps with his brother James, but Josephine felt that this plan would tie them down. "So we compromised," she recalled. They would continue prospecting while scouting for a suitable ranch. They sat and planned like children, drawing straws to decide which point on the compass they would follow to their next destination.

They started out in central Nevada, beginning in Tonopah, where there was news of a big silver strike. In a pattern that they would follow for the next twenty-five years, they traced a ragged circle around the warm winter deserts of California and Arizona, then went west for family visits to San Francisco, and south for Los Angeles summers, following the sun and the mining sensation of the day, never fully giving up the sporting life, but also enjoying the trek for its own sake. Risk was their oxygen, and ennui was a kind of death to be avoided at all costs. Neither of them knew how to live without some new frontier beckoning on the horizon.

THEY OUTFITTED THEMSELVES for the desert in Los Angeles, purchasing a big Studebaker wagon with four heavy horses, as well as a lighter spring wagon with another team of horses. They would drive the lighter wagon, while Al Martin, a horse trainer and partner from their racing days, filled the Studebaker with provisions and a heavy load of what the Tonopah newspaper called "software," otherwise known as liquor, which would stock a new saloon. Wyatt was at his best planning for these trips, as careful about the horses' comfort as he was about theirs. Josephine laughed at her own efforts to smuggle in extra cooking pots and other nonessentials, which Wyatt would cheerfully toss out of the wagon as soon as they hit a rough patch of road. He was more forgiving about the stray dogs and cats that Josephine had a habit of rescuing. Their portable home included a tent, folding chairs, and a mattress, which would be repacked every morning and covered with a waterproof tarpaulin. An excellent cook but an indifferent housekeeper, Josephine ceded the chore of making the bed to Wyatt, who was "the soul of neatness."

Disdainful of the overalls or bloomers worn by some of the women they met on the road, Josephine was increasingly sensitive to her age and weight. Her wardrobe was basic and comfortable, but she still had a horror of "mannish attire." Trousers were particularly unbecoming on a lady "past the first blush of youth," she decided. Her concessions to the rigors of the desert would be shorter dresses, stronger fabrics, stout high-laced boots with low heels, and a wide-brimmed hat or sun bonnet to protect her from the desert wind and sun.

"I recall the days from that time . . . with a greater degree of pleasure than any other portion of our life together," Josephine reminisced to Mabel Cason and Vinnolia Ackerman. Every expedition—and there were dozens of them—renewed her sense of adventure and the intimacy of the road trip.

"[Wyatt Earp and his wife and A. Martin] are good citizens and we welcome them," announced the *Tonopah Bonanza* newspaper. Tonopah did not yet have rail service, and Wyatt had a plan to deliver supplies to the growing town while they scouted a location for a suitable ranch. Once again, his timing was off, though not as badly as it had been in Tombstone; the delivery business was limited. Instead, Wyatt immediately became busy with a new saloon, the Northern, within days of their arrival. Despite a deadly pneumonia epidemic that nearly decimated the camp, Wyatt scored with his usual golden combination of liquor, gambling, and prostitution. Nor was he completely done with police work: he was hired privately by the eminent mining engineer John Hays Hammond to protect mining claims and he also accepted a commission as deputy U.S. marshal. The largest customer for his freighting business was Tasker Oddie, a Brooklyn-born lawyer, general manager of Tonopah Mining Company and future governor and senator of Nevada, who impressed Josephine with his pleasant ways and fine education.

But Wyatt's rapid return to saloon keeping and the sporting life, or what Josephine called "business," upset her. She had fooled herself into thinking that they had left these pursuits behind in Alaska. "I was against him going into business," Josephine protested. "I hated it and I believed he did too . . . but for every man in the prime of his life the clamor and barter of the marketplace has a certain fascination."

She left Wyatt and Tonopah to spend time in Oakland with her sister Hattie, who was about to have a baby. Emil Lehnhardt Jr. was born on May 24, 1902. When she returned, Wyatt had lost interest in Tonopah and was already searching for their next camp. He sold his freighting equipment, retaining one team of horses and their large wagon, and transferred ownership in the Northern Saloon to Al Martin. Prospecting had proved to be lean, but before they left, Josephine and Wyatt filed three claims in Tonopah—naming one of them after Hattie's son.

AFTER A SHORT summer break in Los Angeles, Josephine and Wyatt traveled to Goldfield, Nevada, the latest boomtown. The movements of Wyatt Earp "and his wife, dog, and trusty rifle" from Tonopah were followed by the newspapers. As if Josephine had dictated the press release, one reporter noted that Wyatt Earp had given up saloons and peacekeeping. "He has foresworn the green cloth and the automatic revolver and is now dependent on his ability as a miner for a living as well as his fame."

In Goldfield, they were reunited with Tex Rickard, whose New Northern was the most popular saloon in town. But it was Tex's career as a fight promoter that exploded in Nevada. To raise interest in an upcoming championship fight, he dumped prize money of more than $32,000 in gold pieces into the window of the New Northern. The stunt drew national attention and put Goldfield and Rickard on the map. While the glint of Goldfield's reputation was soon gone, Rickard's reputation soared. He went on to become the founder of Madison Square Garden and the most famous fight promoter in the country, drawing huge crowds and championship purses that reached $1 million.

More Earps showed up in Goldfield. Virgil and Allie arrived in January 1905 from Colton, where Virgil had been running a detective agency. Now there were two couples traveling together. Wyatt and Virgil still looked and thought alike, while their wives could not have been more different. Allie bore a heavy load of grudges against Wyatt that dated back to Tombstone and his desertion of Mattie. Josephine sniffed at Allie's down-home accent and lack of sophistication. Their conflicts must have added some serious spice to the evening meals around the campfire. However, the four of them shared a love of the open road. They prospected as a quartet with more patience than luck until they found promising signs near the Colorado River, the region where Wyatt and Josephine eventually staked out some twelve claims in the foothills of the Little Turtle Mountains. They called it the Happy Day mines, Josephine explained, because it was their favorite place.

Virgil and Allie, however, did not share in the discovery. According to Allie's niece Hildreth Halliwell, Wyatt cut them out deliberately, sending them home with the prediction that "there wasn't any use in staying out there any longer, no mines around there." Halliwell was usually quick to blame Josephine, summing her up as a "clever schemer" who never did anything for anybody without getting paid for it. But in this case, Allie's niece pointed her finger straight at Wyatt.

A rift might have opened up between the brothers had they not been immediately consumed with a far more serious concern when they returned to Goldfield: a pneumonia epidemic sweeping through the town. Virgil was the only one of the four to fall sick. His decline was swift: Virgil died on October 19, 1905, at the age of sixty-two.

To many, Virgil had been held in greater esteem than Wyatt or any of the other Earps. He had led "a more exciting life than comes to the average man," noted his obituary in the *Goldfield News*. Between the reticence of the era and his own taciturn nature, Wyatt said nothing publicly about the loss of his older brother, but according to Josephine, his death stirred the still strong emotions Wyatt had about their life in Tombstone and Morgan's assassination.

Allie had few choices about how and where to live without her lifelong partner. She had no savings, and a few secrets of her own. No one knew, for instance, that she and Virgil had not been ceremonially married, which disqualified her for a pension as a Civil War widow. She accepted the invitation of her niece Hildreth Halliwell to move to Los Angeles. Wyatt and Josephine visited her there, often accepting an invitation to stay for Sunday supper. But without Virgil's warmth to counter Allie's astringent wit and mistrust of Josephine, the three survivors were linked by loyalty and shared history rather than affection.

Of the eight men at the O.K. Corral on October 26, 1881, only Wyatt was still alive.

After Virgil's death, Josephine and Wyatt left Goldfield and resumed their usual pattern. The desert drew them like a magnet. As soon as they unpacked in Los Angeles or another one of their favorite coastal cities—San Diego, San Francisco, or Santa Barbara—they longed to return to their camp near Vidal, California, close to the Arizona border. They had lost none of their love for sleeping under the stars, the smell of rain on the desert, and nights that induced such a profound sense of serenity that Josephine suggested "it would have been hard to find two people with less need for a nerve specialist than we in those years."

They were blessed with good health; in all their decades of camping out, neither of them had consulted a doctor or dentist more than once or twice. The cosmetic effects of outdoor life concerned Josephine more. She reminded Wyatt that "there are not many women who would live as I do, roughing it with you in this hard country." Wyatt took the hint and responded loyally: "There are not many women who keep as well as you do. You never need to send for the doctor. Anyway you look good to me." That was good enough for Josephine, who accepted his plain-speaking words in the place of any high-flying romantic compliments. His days of buying her jewelry might be over, but he was still her hero and protector. One of her happiest days in the desert was when Wyatt surprised her by building a tree house, which he had designed as a bedroom that would keep her safe from snakes.

They had been a strikingly handsome couple in their early days together, but Josephine admitted ruefully that Wyatt had aged far more gracefully. At nearly sixty, his years only added gravitas to his wiry frame, chiseled features, and charisma. Discriminating young women such as Grace Spolidoro, daughter of Wyatt's close friend Charlie Welsh, recalled Wyatt as "tall and erect with steely blue eyes that could stare right through you." Neighbors remembered his deep voice, carefully chosen words, and chivalrous manners. When in town, Wyatt never joined the breakfast table until he was fully dressed in his dark suit, his collar fastened, necktie in place.

The towering, still powerful figure of Wyatt Earp fit the rugged saga of Tombstone, while Josephine was looking more like a frumpy old aunt. Gravity and extra pounds had taken their toll on her hourglass figure, and her beautiful hair and skin showed the effects of their nomadic circuit. With no particular need to dress up, she wore shapeless housedresses that fit their life at the camp but did little to remind anyone of the saucy young woman of Tombstone. Age did nothing to diminish her bright eyes or the gift of mimicry that made her popular with children, as did the sweets she always had ready to share from her sister's candy store, Lehnhardt's of Oakland. To teenage Grace Welsh Spolidoro, Josephine appeared to be merely a "good guard," plump and bosomy, always fussing with her hair and hat, inexplicably spoiled by Wyatt. She was amusing, Grace recalled, though she also criticized Josephine for her preference for gambling rather than taking proper care of Wyatt.

They lived frugally, but their Alaskan stake was almost gone. At one point, Josephine considered a return trip that might replenish their fortune. "We are talking some of going to Alaska for the summer," she wrote to a friend, after she and Wyatt were invited to join a prospecting trip. "And I am just in for it, as I do want to find a good gold mine. And I think that's the country."

Wyatt was then sixty-six, and she was fifty-four. Realistic planning for the future had never been their strong suit.

JOSEPHINE WAS IN the desert on April 18, 1906, when San Francisco and the Bay Area suffered a major earthquake, followed by a fire that consumed most of the city over the next three days. In what has been described as

America's greatest urban catastrophe, more than 3,500 people died, and half of the city's survivors were left homeless. Business came to a total halt. The immigrant Jewish quarter south of Market Street, where the Marcus family once lived, was devastated. So was the personal history of several generations as the earthquake opened a yawning chasm in pre-1906 history that would never be filled, as archives, photographs, public records, and personal memorabilia disappeared forever.

It was a time of communal suffering. The Lehnhardt mansion in nearby Oakland was spared and became even more the center of the family as Hattie and Emil made room for displaced friends and family; Sophia Marcus was already living with them, and Rebecca and Aaron followed with their children. Josephine and Wyatt were the last to move in.

The next few years brought severe aftershocks of a more personal kind. Forty-eight-year-old Nathan Marcus, only recently returned from Alaska, died soon after the earthquake, though the cause of death was listed as diabetes. He was buried next to his father. Nicholas Earp outlived two of his three wives and five of his nine children, and died at ninety-four in 1907 in the Soldier's Home near Los Angeles, identified in his obituary as "father of the noted Earp boys." Lucky Baldwin died on his ranch in 1909, a still vigorous eighty-year-old who left behind a diminished but still substantial fortune in California real estate.

Times were changing. The only constant was Tombstone, constantly lurking in the background, threatening to reveal the secrets of the past.

"OF COURSE YOU know all about the trouble Mr. Earp got mixed up in," Josephine wrote on August 2, 1911. She and Wyatt had moved from San Francisco to Los Angeles, and Wyatt was arrested in a so-called bunco scheme (con game, in today's parlance) on July 21, 1911. A man who initially put up the money for a friendly game of faro had tipped off the police when he realized that he was the intended mark.

As if to underscore the bewildering changes that twentieth-century life was presenting to an aging Wyatt Earp, the man who had made his living as a frontier gambler was arrested for running a card game in a big-city hotel. He tried to hide his identity at first, but his alias was quickly pierced, and then the other three perpetrators were all but forgotten: it was Wyatt's name that made the headlines. Reporters linked the arrest back to Wyatt's record as controversial Arizona marshal and referee. Thirty years after the O.K. Corral, and fifteen years after the Sharkey-Fitzsimmons debacle, the national headlines proclaimed, "Notorious Marshal Who Disqualified Fitzsimmons Arrested in Raid." Wyatt was saved from a possible jail sentence only because the police bungled the case: they had instigated the raid before the game had actually begun. Wyatt and his associates were released.

Josephine was miserable over Wyatt's arrest, but stalwart in his defense, insisting "Everything was put on to him just because he was Wyatt Earp." For help in this crisis, she turned to John Flood, a young engineering graduate of Bucknell and Yale who had been introduced to the Earps by a Los Angeles neighbor. With no family of his own, the unmarried Flood admired Wyatt like a devoted son. His relationship with Josephine was less reverent, but for Wyatt's sake Flood hid his irritation with the often bossy Mrs. Earp. Flood became their legal and financial adviser, real estate agent, and engineering consultant. He was their secretary, typing their letters, sometimes taking dictation from Josephine. Efficient and organized, he kept careful records and carbon copies of their correspondence.

If Wyatt had intended the card game to help their financial situation, he had made it even worse. The couple had to hire a lawyer and use their property as collateral. But in the court of public opinion, Wyatt was well defended by his influential and articulate supporters, the posse of old friends who came to his aid with their words instead of their guns. In former days, Wyatt himself had coordinated the defense. Now, it was Josephine and Flood protecting him and orchestrating these outpourings of support. "Mr. Earp knows his friends believe in him. And for strangers, he cares not," she declared stoutly to Flood after Wyatt was released from jail.

Josephine foresaw the need for greater vigilance in managing Wyatt. "This will teach him a lesson never again to enter a gaming club," she hoped. She wanted him to be known as a gentleman who owned real estate, horses, and mines. A day at the track, wearing fine clothes, and drinking champagne with Lucky Baldwin: in her view, that's where Mr. and Mrs. Earp belonged.

So she wanted to believe—except that Lucky Baldwin was dead, and Josephine could no longer fit into her beautiful gowns. Wyatt had nearly found himself back in the "cold and silent calaboose."

What was commonplace at the frontier had been pushed to the margins of American society. Wyatt was caught on the wrong side of the pendulum that swung back and forth between taboo and tolerance. Gambling tables were forbidden, and now an even bigger change was coming: Prohibition. Wyatt had owned or operated ten saloons in six towns in six different states. He and other saloonkeepers had been more than convivial alcohol sales-men; they had played essential roles in frontier communities as bankers, information sources, and political conveners, furnishing the government with much needed tax revenues. Now the crazy quilt of state regulations about alcohol distribution was being gathered together into an unstoppable national movement.

When Prohibition was proposed in Arizona in 1909, the *Tombstone Epitaph* editorialized: "If the present and proposed prohibition laws and antigambling laws become effective in Arizona, Tombstone will become as prosaic and goody good as a New England church town. Shades of Holliday and Earp, how times have changed."

Remarkably, Tombstone itself would soon be dry, an early casualty of state-enacted prohibition six years before the rest of the country closed its bars and saloons.

. . . .

JOSEPHINE HAD ALWAYS been the unpredictable one in the Marcus family, the one who seemed to invite the question, "What will happen to her next?" It was her sister Hattie, however, who suddenly eclipsed even Wyatt's arrest and the advent of prohibition.

"Rich, but Weary of Life, Confectioner Kills Self" was the unexpected headline on January 27, 1912. Hattie's husband, Emil Lehnhardt, was found dead in the basement of his store, a self-inflicted bullet wound in his head. No one had gauged the depths of Emil's depression or even knew that he had a gun. Known to all as "Oakland's Candy King," a pillar of the community, Unitarian church, and Masonic order, he personified the self-made man, a first-generation American who had arrived in San Francisco as a traveling salesman and built Lehnhardt's from a local sweet shop into one of the largest confectionary businesses on the Pacific coast, employing 140 people. He was about to open another factory, this one primarily for exporting to Pacific markets.

Lehnhardt had recently lost his sister after a long illness that left her paralyzed. After he too experienced a series of minor strokes, the doctors warned that he might be similarly afflicted. He accepted his doctors' recommendation to retire and authorized Hattie to sign checks. Son-in-law Estes Joseph Cowing was recruited to become general manager of the business. Cowing had married Edna Lehnhardt in October 1908, and they had two children, Emil and Marjorie.

Hattie and Josephine were still struggling to regain their balance from Emil's suicide when their mother, Sophia Marcus, died on August 19, 1912. She had been living in Hattie's large home since the death of Josephine's father. Sophia was identified as a "pioneer of the state" in the obituaries. Despite her dramatically different path, Josephine had always remained close to her mother, who inspired Josephine's love of nature. It was her mother's lullaby, "Lieber Gott," which haunted Josephine in her first lonely days as a young runaway with Pauline Markham's Pinafore troupe. Now there were three members of the Marcus family in the Hills of Eternity cemetery. Both Josephine and Wyatt had lost their parents, another sign that their age was catching up with them.

For more than five years, Hattie and her son-in-law ran the business together but then a third family cataclysm occurred when Edna got divorced. An aspiring artist, Edna moved back into the family home at Oakland with her children. Hattie took over Lehnhardt's and became a well-known businesswoman of the Bay Area, featured in the society columns as a philanthropist and supporter of the San Francisco Opera Association and other cultural organizations, and listed in *Who's Who* among the women of California.

Josephine traveled continually to Oakland during these trying years. "I am particularly glad to know that Edna is well again after her sorrow and trouble," John Flood wrote sympathetically.

It was during this time of close family connection that Josephine confided her money problems to Hattie, who offered her sister a monthly subsidy, which she increased as Wyatt's earnings declined. Hattie became a silent partner in some of the Earps' mining claims and oil wells. Edna sent gifts of clothing to Josephine. In 1920, Wyatt began to explore the oil fields in the Kern River area near Bakersfield, California. He filed for permits in his name, but later transferred these investments to Hattie, with the understanding that Josephine would be entitled to receive 20 percent of any income derived from the oil wells. Wyatt was paid a management fee, which probably helped to soothe his pride in accepting money from Josephine's family.

It was unclear what the future might hold. The wells proved to be profitable, but no one knew how long that would continue. Hattie was younger than Josephine, and both were younger than Wyatt. Presumably, Hattie or her heirs would continue to support Josephine with the income from Wyatt's wells.

Hattie became the lady that Josephine was not: rich, socially prominent, and sophisticated. Josephine was capable of admiring her sister and relishing the way she lived, while being fiercely loyal to Wyatt and the choices she had made as Mrs. Earp. Still, it was a relationship that occasionally irked Wyatt; it reminded him of his financial dependence on Hattie and also took Josephine occasionally away from him to destinations he could not afford, or no longer had an interest in visiting. Hattie often invited Josephine to accompany her to San Francisco events, and they traveled together to New York and Boston in Hattie's own drawing-room train car. "You know I was born in New York City and I am so happy to think I can go and see the wonderful city," she wrote to Flood, adding with some defiance: "I wrote and told Wyatt don't know how he will like it. I am going just the same as it will be a wonderful trip for me." It was one of her happiest days when her sister threw a grand party to celebrate Josephine's sixty-fifth birthday on June 2, 1925. "That was a grand luncheon at the Hotel Oakland," John Flood wrote, after receiving clippings from Josephine about the impressive guest list and elaborate menu. "What a society lady you are; that is right, don't miss anything." Presumably from Hattie, Josephine received a birthday gift of sixteen crisp $1 bills.

Their sister Rebecca stayed the closest to the Marcus family roots, and occasionally hosted Jewish celebrations attended by Josephine and Hattie. The three petite women were a handsome trio. Josephine and Wyatt attended at least one Passover celebration in San Francisco, probably at Rebecca's home. The sight of Wyatt Earp putting down his gun and donning a *yarmulke* at a seder was never to be forgotten: many years later, one guest told actor Henry Fonda about his impressions of that memorable evening, taking the opportunity to compliment Fonda on having captured Earp so well in *My Darling Clementine.*

Josephine entertained her sisters' children and grandchildren for long stays at their desert camp and visits to Coronado Island. They enjoyed her company, while Wyatt, they agreed, was simply easier to adore. They often chafed at what they called Aunt Josie's "suspicious nature." As their temporary guardian, Josephine was more of a disciplinarian than Wyatt, ready with a stern lecture about the dangers of smoking and dancing. "Coronado had a ferry [to San Diego] in those days," Edna's daughter recalled. "We were taking the ferry to go to a movie. Instead of that, we went to three! They were westerns! When we came back, we got off the ferry and Aunt Josephine was standing there on the pier yelling and screaming. She was furious. Oh, she was so mad because we had left early and now it was nine at night." On the other hand, "Uncle Wyatt wasn't a bit upset about it, he just ignored the whole thing." Josephine would not be the first grown-up to have regretted—or forgotten—her own wild days, and was now finding it convenient to condemn all those things she did when she was young.

THEY WERE AGING with reasonable grace. If only the devils of Tombstone would stay in the past, Josephine believed she could be content.

Instead, the stories about Wyatt and his brothers never stopped swirling around, occasionally fatuous and fawning, frequently inaccurate, and most often pointing directly back to the O.K. Corral. Before he died in 1912, Johnny Behan was interviewed and portrayed himself as the hero who defended Tombstone against the villainous and violent Earps. Josephine and John Flood recruited Wyatt's friends to correct factual errors and serve as character references. Remind people that it was the "better element" that supported the Earps, people who later became important members of society, as opposed to the cowboys, who died in jail or gunfights, Josephine coached their friends. Her willing army of spokesmen included John Clum and William Hunsaker and George Parsons from Tombstone, as well as younger people of influence such as Tasker Oddie and John Hays Hammond. When the *Los Angeles Herald* wrote about the Earps as "bad men," George Parsons countered that Wyatt was "a benefit and a protection to the community he once lived in." He did note the Vendetta Ride as a "notable exception," but suggested that after the shooting of one brother and the killing of another, anyone would have done what Wyatt did.

Until his death in 1921, Bat Masterson was Wyatt's most effective spokesman, a gifted writer who could always be counted on for a colorful quote. The public's view of Wyatt Earp was greatly shaped by Masterson's early descriptions of Wyatt as "a very quiet man, but a terror in action, either with his fists or with his gun." In a 1910 interview in the *New York Herald* he defined Wyatt's distinguishing trait: "More than any man I have ever known he was devoid of physical fear. He feared the opinion of no one but himself, and his self-respect was his creed."

That Wyatt made his living as a gambler was another issue: on this point, both the Earps were vulnerable, though in different ways. In a letter that Flood wrote under his own name, he took his cue from Josephine's oft-expressed argument about the changing times: "That the Earps were gamblers is not for me to question. It seems to have been a part of everyday life on the frontier, during the pioneer period. . . . Times have changed; people think differently. What were established business yesterday are considered vices today, and have been outlawed."

Flood wrote another point-by-point rebuttal of "Lurid Trails Are Left by Olden-time Bandits," a particularly inaccurate article that appeared in the *Los Angeles Times* under the byline of J. M. Scanland. No, the Earps had not been driven out of Dodge City by Bat Masterson—Wyatt had served as marshal there with distinction, and he and Bat were lifelong friends. It was Virgil who was the chief of police in Tombstone, not Wyatt. Virgil did not kill Frank Stilwell; Virgil was lying injured in the train that stood waiting in the Tucson station. There was no substance behind the accusation that Wyatt was the mastermind of a bandit gang; he had the confidence of Tombstone's mayor, the Wells Fargo Company, and the town's leading businessmen and respectable ranchers.

And, no, Wyatt Earp was not dead.

"I trust that what I have written, meets with your approval. . . . I could not remain silent after reading the article as it appeared in the *Times*," Flood wrote to Josephine, who was in their desert camp at Vidal and had not seen the newspaper yet. Josephine was pleased with Flood's spirited defense. Then she picked up the telephone herself and gave the managing editor a piece of her mind: "I told him I wanted to have every untruth corrected and printed in the same sensational manner that [the original article] was printed."

In the case of the *Los Angeles Times*, Josephine won a major victory—a public retraction. The combined efforts of Wyatt's friends and her own call to the managing editor, as well as the incontrovertible fact of Wyatt's being very much alive, elicited a prominent correction notice under the satisfying headline, "Earps Were Always for Law and Order—Writer's Statements Result in Real Facts Being Given Notice." Readers were told that it was all an unfortunate misunderstanding by a writer. "Through information furnished by relatives of the Earps," the *Times* now knew that the Earp brothers had "nothing in common with the bandit gangs, but that, instead, they did everything in their power to protect the people and to uphold law and order." In fact, the *Los Angeles Times* went so far as to repeat the assertion of relatives that "none of the Earp brothers ever opened a saloon or a gambling house."

If there was irony in the *Times* retraction, Josephine cared not. She had a strategy for public relations: rapid response, rebuttal with carefully sculpted facts, and character references delivered by influential friends, all topped off with her personal demand for a withdrawal. She brought a modern sense of celebrity to the task of shaping Wyatt's image to contemporary standards and away from his past as a gambler, saloonkeeper, gunfighter, pimp, and womanizer.

She put her plan into effect for the next offending article, this one from a New Jersey newspaper owned by Hearst. She asked Wyatt's friends to reprise their previous letters, and for the personal touch, she commissioned Flood to type a letter to William Randolph Hearst, son of the senator who had traveled under Wyatt's protection decades before. "You know just what kind of a letter I want," she told Flood. Despite her agitation over the constant barrage of criticism, she lost none of her zest for battle, emphasizing "this time we will *fight them all*." She wrote in the stream-of-consciousness style that overtook her when she was excited: "Please say that in the early days of Tombstone his father George Hearst came to Wyatt and his brother Virgil asking for protection from some of the toughs in Tombstone as he was going out to look at a mine and Wyatt took him out on horseback and stayed with him for 2 days for which after he returned to San Francisco." And don't forget the nice watch that George Hearst sent to Wyatt, she added in a postscript, though the watch had been lost long ago to a Mexican pickpocket.

JOSEPHINE AND WYATT had followed the boomtown circuit since they left Tombstone, always alert to that mysterious signal that was inaudible to mere mortals, the one that whispered of high adventure and glittering treasure. They heard it next coming from their own backyard in Los Angeles.

Hollywood was the next frontier. Josephine believed that it could deliver the dual advantages of making money and burnishing Wyatt's image. Amplified and sweetened, his story could now reach into nearly every corner of the world. Surely it would be strong enough to drown out their enemies.

HOLLYWOOD'S PROXIMITY TO Los Angeles was accidental; early founders like Cecil B. DeMille were looking for a temperate climate and low property values. DeMille considered Flagstaff, Arizona, but settled on Los Angeles, where it never snowed and real estate was an affordable fifty dollars an acre. From a barn on Selma and Vine, DeMille filmed *The Squaw Man*, the first feature-length Hollywood film, on December 29, 1913. The setting was no accident. Western stories and novels tapped into a wave of nostalgia for the frontier. Wild Bill and Buffalo Bill reenactments still drew large audiences. "Cowboys" were in ample supply in the Los Angeles stockyards, where trail drivers from Arizona, Utah, and Nevada congregated after delivering their herds. The better looking of the men were recruited to the studios, together with their own horses and saddles, amazed that "someone would pay them money just to ride around while someone else took pictures of them."

Westerns were also a popular subject because a few of the legends were still walking around Los Angeles. Wyatt Earp and his contemporaries were about to undergo the weirdly modernist experience of moving from the real to the based-on, could-have-been version of their own lives. Having crossed the country in a wagon train, hunted buffalo, experienced Indian raids, gunfights, and the rise and fall of the boomtowns, the still living frontier men and women were becoming fictional characters.

Hollywood would tell many stories, and would take advantage of sound, color, and the latest technology, but Westerns would always be on the menu. Important actors played frontier lawmen and cowboys, and the rest just admired them. "You're the bloke from Arizona, aren't you?" inquired Charlie Chaplin, when he was introduced to Earp at a lunch with director Raoul Walsh and writer Jack London. "Tamed the baddies, huh?" Now into his 70s, Wyatt Earp looked the part, still handsome and straight as an arrow. The same articles and books that harped annoyingly on Tombstone and Dodge City had made him famous and drew some of the biggest Western stars to the modest bungalow he shared with Josephine. Directors such as John Ford, Alan Dwan, and Raoul Walsh sought him out for authenticity, patterning their language and appearance after his and inviting him to their sets.

Josephine often accompanied Wyatt to the movies at the Rialto, one of the theaters owned by their old Alaska comrade Sidney Grauman. She loved the movies, and appreciated the obeisance of Hollywood's aristocracy toward Wyatt, though they had little interest in her. However, as the one who worried most about filling the family coffers, she resented the exploitation of the Hollywood Westerns. Everyone seemed to be making money off Wyatt Earp except for Wyatt, she often complained. Hollywood "consultants" were not yet the norm. Wyatt probably could have made some money as an extra, but the one time he tried, director Allan Dwan reported that Wyatt seemed uncomfortable in front of the camera and disdainful of the make-believe antics of actors such as Douglas Fairbanks. Wyatt's experience with Hollywood was much like his experience with reporters. "After he discovered that they paid no attention to what he told them Wyatt's sly sense of humor was directed toward the movie people," Josephine recalled. "He pulled their legs, telling them the sort of improbable things found in Western fiction stories. To his amazement, they swallowed these tall tales hook, line, and sinker, but were always skeptical of the truth. At this point, my husband gave up in disgust, refusing to have anything further to do with those 'damn fool dudes,' as he called them."

It was the rare actor who brought authenticity to the screen; one exception was Tom Mix, a former ranch hand from Oklahoma. The "King of the Cowboys," Mix made over 160 films and was a frequent visitor when Wyatt and Josephine were in Los Angeles, sometimes accompanying Wyatt to the racetrack. William S. Hart was another Earp acolyte. Where Tom Mix was a rough-and-ready showman, Hart was classically trained, as comfortable in a Shakespearean tragedy as he was in a Western. He looked the most like Earp, according to director Raoul Walsh: Hart was "stone-faced, steely-eyed, was law and order behind the hammers of his six guns . . . projecting the same cool, efficient authority of Tombstone's renowned marshal."

In a series of films, Hart aimed for a gritty portrayal of the West, with an emphasis on realism that endeared him to Wyatt. He had a special interest in the O.K. Corral that began even before they met in person. "It haunted his mind," Hart's son recalled. He took lessons from Wyatt in the fast draw, and proudly displayed a whip that Wyatt had given him. A popular writer and savvy businessman, Hart mailed a thousand pictures weekly to his fans. His children's books, based on the adventures of "Bill Hart's Pinto Pony," were sold in bookstores and, at Hart's expense, distributed free to public libraries.

Hart became an important adviser to Josephine and Wyatt at a time when Hollywood was getting ready to invade Tombstone. Josephine was used to fighting public relations skirmishes, but this battle called for more sophisticated weapons and strategy.

The story of the O.K. Corral was being shaped for the modern era and new media. To her relief, there were no parts yet for the young Josephine Marcus, and the role of Johnny Behan was downplayed or diffused into different characters. Wyatt Earp, however, was front and center. It was Wyatt who emerged unscathed from the first hail of bullets, Wyatt who caught the fallen Virgil and Morgan in his arms, and Wyatt who led the Vendetta Ride. Lawman or vigilante? Avenging angel or ambitious politico? Peacekeeper or stagecoach robber? These were the alternative roles that would determine Wyatt's legacy. Tombstone was inescapable. The benign labels that Josephine had cultivated for Wyatt—capitalist, prospector, businessman—had little chance against a dramatic lone figure who counted kills in notches on a gun barrel. She would have to work harder to salvage his reputation within the tenacious tale of Tombstone.

As the old-timers died off, the memoirs of the remaining eyewitnesses became all the more precious. Josephine was convinced that Wyatt's biography would be the most compelling best seller of all. The story was there: all they needed was a writer. Wyatt resisted mightily. He deflected most inquiries, even from friends and family, with a question of his own: "Isn't there something more pleasant we could talk about?" Josephine's nieces could not imagine him as a gunfighter in the "wild and wooly west." When his nephew Bill Miller pumped him for stories about the Vendetta Ride, Wyatt would only admit to killing Frank Stilwell, which he claimed with pride.

Josephine's determination gradually softened Wyatt's opposition. He was concerned about his health and finances. Defensive strategies and counterattacks could take them only so far, especially since the passing of some of Wyatt's most stalwart defenders, like Bat Masterson. Since books and movies were obviously going to be made, she argued, they should take an active role in shaping them—and should share in the profits. She wore down his objections. "I have never put forth any effort to check the tales that have been published. . . . Not one of them is correct . . . what actually occurred at Tombstone is only a matter of weeks. My friends have urged that I make this known on printed sheet. Perhaps I shall," he concluded. "It will correct many mythic tales."

Some writers were politely rejected. An early offer came from novelist Forrestine Hooker, daughter-in-law of the prominent Arizona rancher Henry Clay Hooker, who had sheltered Wyatt's posse and defied Behan's orders back in Tombstone. Hooker drafted a manuscript that she showed to Wyatt, and although she was clearly his advocate, Wyatt turned her away. His refusal was less about literary discernment or doubt of a woman's ability to tell his story than about his lingering reluctance to stir up more notoriety. Another aspirant was Dr. Frank Lockwood, Dean of the University of Arizona. Wyatt refused again, though he and Josephine were impressed with Lockwood's academic credentials and granted him an interview.

Surely they had the best partners right under their nose, Josephine suggested: their trusted friends John Flood and William S. Hart. Flood already served as their public voice, writing letters that Josephine commissioned, often with specific language and approaches discussed by Josephine and Wyatt, typed by Flood, and signed by Josephine under Wyatt's name. It was an unusual three-way collaboration, but it worked. Their personal and financial lives were intertwined, as Flood became their adviser in the desert mining and real estate interests that still anchored their cycle of Los Angeles summers and desert winters.

Josephine hoped that Wyatt's biography would be the basis for a William S. Hart film. In July 1923 Wyatt (i.e., Josephine and Flood) wrote to Hart with a formal proposal: "During the past few years, many wrong impressions of the early days of Tombstone and myself have been created by writers who are not informed correctly, and this has caused me a concern which I feel deeply. You know, I realize that I am not going to live to the age of Methuselah, and any wrong impression, I want made right before I go away." Hart would be the "mastermind," the producer and director as well as the star who would portray the real Wyatt Earp. With Hart's blessing, and the promise of access to Wyatt and Josephine, Flood set to work.

He had already spent nearly twenty years with the Earps, but Wyatt had been as reticent about discussing Tombstone with him as he had with everybody else. Josephine knew some details gleaned from Bat Masterson and from snatches of conversation around the desert campfires in the early days of prospecting with Wyatt and his brothers. For the rest, Flood turned to some of the old-timers. Sol Israel, now an agent for a California vintner selling alcohol for medicinal use, was particularly helpful and eager to support his old friend. John Clum and others agreed to be interviewed.

Surely the best source would be Wyatt himself. But Flood found his idol and friend of many years, now his subject, to be the man described by Bat Masterson: preternaturally calm, steely-eyed, intimidating—and laconic.

At Wyatt's request, the men met alone, usually in Los Angeles, sometimes for a few hours and sometimes for an entire afternoon of probing questions and terse responses. Flood dated each of their interviews, taking extensive notes that he transcribed into a detailed outline and shaped into a draft. Wyatt's memory was remarkable. Now deep into his seventies, he recalled names, dates, locations, and physical descriptions, even names of children. They spent hours rehashing the actual gunfight. As Wyatt spoke, Flood prepared a detailed diagram that placed each participant on the specific spot he occupied that day, and captured Wyatt's telegraphic style in his notes: "Doc Holliday was about 5'10 ½", slender, good looking, died Colorado Springs—order of the Odd Fellows," he wrote. "Morgan looked most like Wyatt. Little lighter than Wyatt. Handsome fine complexion. [Tombstone mining executive] E P Gage gave Morgan Earp a horse, mistaking him for Wyatt . . . Curly Bill, about 5'11 ½" straight, looked like a Cherokee Indian, heavy build, 180 lb., was solid."

Josephine was Flood's first reader and the one who kept an eye on the clock, mindful of Wyatt's fatigue. She kept Hart engaged through lively letters under Wyatt's signature that promised that "the script which I am having written will be ready in a short time." However, progress was slow. Flood had the difficult task of portraying Wyatt Earp while adhering to Josephine's vague ideas about domesticated adventure and romance. She and Flood had endless discussions about what to put in and what to leave out, which she followed up in cryptic orders ("you know what I mean"). Although an outsider could hardly have understood her references, she urged him to throw away her letters immediately after reading, and to seal his envelopes with wax.

A clean story full of true facts, pep, and romance. Those were Josephine's constant refrains to Flood while he was writing throughout 1925 and 1926.

Josephine realized that they were in trouble almost immediately. Flood was simply not a writer. Although he alone had special access to Wyatt, it was hard to pull anything but the bare details from a subject who preferred a sharp glance full of meaning to a fulsome paragraph. Nor was Josephine a skillful editor. Her instincts were good: she knew something was wrong, but had no idea how to fix it. Flood's writing was wooden even when narrating the most exciting parts of Wyatt's life. So she pressed her suggestions again and again in her staccato style—write about Wyatt's early days driving teams from San Bernardino to Salt Lake City, his bravery in Dodge City, his rescue of a young girl and her mother from a Wichita fire, Rex Beach playing the banjo for the dancers in Rampart. She volunteered the services of her well-educated nieces. She passed along advice from Hart and from one of Jack London's daughters and from one of her cousins. In fact, she consulted everyone she knew, and all of them seemed to have something worthy to propose. Hollywood was waiting, she reminded him. "Just think John H. Flood pop up on the screen." When glory and bribes did not work, she tried guilt and threats. Sometimes she sounded like a mother scolding her young son: "So hurry up now like a good boy and finish it up as soon as possible," adding an even more humiliating postscript: "Now be a good boy and do as I ask of you, for just this one time. And I will buy you a sucker."

At least she could be sure that Mattie and young Josephine were written out of Flood's manuscript completely. Her instructions to Flood were all in the third person, always about what "he" or "they" did, as if she had not been there herself. Her reticence was not limited to Tombstone. "They had a gay old time getting up to Alaska," she reminisced, but it was only Wyatt's bold actions that she wanted him to include. "Cut out you know what—I mean as we want a good clean story and one that will pass too and yet be full of the truth and lots of pep in it too." It is not clear how much Flood actually knew about the censored matters: the women upstairs at the Dexter, the story of Mattie Blaylock, Josephine's year as Mrs. Behan, perhaps a simmering scandal or two from Wyatt's popularity with the women of Harqua Hala, San Francisco, and Nome.

A clean book also meant a very dry Wyatt Earp.

Even though alcohol was not illegal during Wyatt's heyday, it was now, and no one knew when Prohibition would end. Josephine needed to mold a Wyatt Earp who complied not only with the laws of the 1880s but also with the laws of the 1920s. What sort of lawman would Wyatt Earp be if he broke the law, even laws that did not exist yesterday?

Thus was born the notion that Wyatt hardly drank a drop. Wyatt does appear to have been somewhat abstemious, but he was no teetotaler. Wyatt's friend, Charlie Welsh, who always kept a flask of wine or whiskey on his dining-room sideboard, was known to disappear regularly for days "to see property," the family euphemism for a drinking binge, and Wyatt was his chosen companion for some of these trips.

No alcohol and no Josephine, but Wyatt's story should really have romance, Josephine insisted. At least love had not been outlawed, though social mores had certainly changed since Tombstone. "There must be a woman in it," Josephine emphasized. After all, the book was intended as the basis for a future script, and "you know how Hart likes to make love," she teased Flood.

In fact, Josephine was looking over their past as if it had been a script. Wyatt must always have the starring role, and her own character would be absent or scattered into surrogates.

Just once, in a remarkable 1925 letter, Josephine lowered her guard. She suggested that Flood explain it was a woman who triggered the personal animosity between Wyatt and Johnny Behan. This letter was the closest Josephine ever came to telling the truth about her role at the O.K. Corral.

"Call her Ann Ellen," she coaxed Flood. "Quite a pretty girl age 19 years," her pseudonymous Ann Ellen would be "in awe of Wyatt," and even though Sheriff Behan was in love with her, "she really did like Wyatt and Behan was very jealous and that of course made it hard for Wyatt too." Perhaps because of the potentially explosive connection between "Ann Ellen" and young Josephine Marcus, this story never made it into Flood's final version, nor did Josephine ever allude again to the story of Ann Ellen.

The faithful but humorless Flood stuck to the basics. The resulting manuscript was dry as dust.

"AT LAST! THE story is finished," Flood declared early in 1925. But it was not. For another year, Flood continued to interview Wyatt and draft new versions under the pressure of Josephine's latest brainstorm. Although she constantly urged him to finish quickly "for Wyatt's sake, for my sake," she suggested that he expand his scope to include Wyatt's boyhood and later Alaskan adventures. Flood wanted to concentrate primarily on the years in Kansas and Arizona. It would take far longer to produce a complete biography, but he was resigned to the necessity of following Josephine's direction. "A romance might be the best thing. I don't know. I will have to think it over," he muttered aloud.

The deadline kept slipping. Josephine worried that Bill Hart would lose interest. Still more ominously, Wyatt was growing weaker. Her letters trace his lack of appetite and diminishing strength. "I am afraid he is not long for this world," she fretted.

Although much younger, Flood too was in poor health. He experienced eyestrain so severe that he was bedridden and blindfolded for two days. With almost sadistic timing, Josephine chose this time to send him one of her longest letters, written in cramped handwriting on pale blue note paper. "Well, Mr. Flood, I am afraid this letter will be very hard on your eyes. So don't you try to read it all at one time," she ended with a hollow warning.

Although she was undoubtedly annoying, Josephine's meddling was not responsible for the fact that Flood was mired in a badly jumbled, cliché-ridden, boring manuscript. He insisted that he was merely taking dictation from Wyatt, and the dull narrative does sometimes have all the grace of a court deposition. Where Flood reaches for dramatic effect, his subject comes across as George Washington in a religious allegory, as in his description of Wyatt's birth: "And the morning and the evening were one, and the infant, Wyatt Berry Stapp Earp, slumbered in the Land of Nod." Wyatt is referred to as "the young plainsman." Flood was particularly inept at psychological portrayals: "Earp could feel the warmth of the conspirator's body as he leaned against him; the pulsations bent against his own and then there was a throb; something that felt like nerves, and the tenseness of muscles at the drawing of a gun." He did, however, remain faithful to Josephine's emphasis on a clean story. He decorously drew the curtain across any romantic entanglements for Wyatt or any of the Earp brothers. In fact, all the Mrs. Earps had vanished, except for one allusion to an unnamed wife at the end of the manuscript, preparing the evening meal at the desert camp while Wyatt Earp goes out to "stake his last claim."

Flood finally did finish. He sent the manuscript to Hart with a plea for mercy to a first-time author. Hart sized up the manuscript as clumsy but thought it had enough promise to send it to several publishers and to the *Saturday Evening Post* in the fall of 1926, with a generous endorsement by William S. Hart and the promise of one "never before seen photograph of Mr. Earp."

WHILE FLOOD WAS struggling through his task, other writers continued to approach Josephine and Wyatt. Walter Noble Burns, a Chicago journalist who wrote *The Saga of Billy the Kid*, was eager to follow up that success with another dramatic Western. He already had a broad readership, and his book had been sold to Hollywood; Flood tried to get a copy of *Billy the Kid* at the library and was put on a waiting list. Wyatt granted Burns an interview, but made it clear that he had already selected his biographer.

Josephine realized with consternation that Burns would have been a better choice than Flood. However, she had to agree with Wyatt that loyalty was everything, and they must stick with Flood. "LUCK BUT NOT FOR US," she fumed. Burns reluctantly accepted Wyatt's decision and suggested that instead he would write about Doc Holliday. Wyatt enthusiastically responded to an opportunity to set the record straight about his friend, and they began a correspondence.

Wyatt, Josephine, and Flood failed to realize that Burns was only pretending to be writing a biography of Doc Holliday. He had never abandoned his first plan. It was not Doc at the center of the book that Burns would call *Tombstone: An Iliad of the Southwest*, but Wyatt Earp.

JOSEPHINE'S WORST FEARS about Flood's plodding manuscript were confirmed by quick, definitive rejections from three publishers and the *Saturday Evening Post*. The reviews were harsh, especially one that referred Flood to a lively new story about Wyatt Earp that was about to be published by Doubleday Page and Company—written by Walter

Noble Burns. With words like *stilted*, *florid*, and *pompous* ringing in his ears, Flood was the first to suspect Burns's double dealing.

Although he had not yet read Burns's book, Flood assumed the worst, and advised Josephine to consult with Hart about how to stop its publication; at the least, she had to make certain that the title would not bear Wyatt Earp's name. "Strike quickly, and hard, and carry the thing to the bitter end," thundered Flood. "They are using something which belongs to you, which is your property," agreed Hart, who recommended a copyright lawyer. The result was a cease-and-desist letter that elicited polite assurance from the publisher that Burns's book would be complimentary to Wyatt, and a follow-up visit by a Doubleday agent, who offered the Earps a 10 percent royalty. Josephine and Wyatt countered with a demand for 50 percent, plus a long list of corrections. The negotiations broke off with Josephine telling the Doubleday agent that Burns was foxy, sneaky, tricky, phony, and low-down, a thief, a wolf, and a liar. The controversy caused the *Saturday Evening Post* to withdraw its offer to Burns for serial rights.

Between the ineptitude of Flood and the duplicity of Burns, Josephine and Wyatt were stymied. At seventy-nine, Wyatt declared, "They have told so many lies that it makes a man feel like putting on his fighting clothes." Josephine's letters about Burns have a frantic, repetitious tone, and an unpunctuated wildness. She feared that Burns's book would be taken as "the real story of Wyatt Earp" and would destroy any possibility of subsequent books. "For once the public had read one story of Mr. Earp, they would not be interested to read the same thing again by another author," agreed Flood. "Immediately then, Mr. Burns would have his story filmed, and there you are!"

But it was Flood who calmed down first. Risking Josephine's wrath, Flood remained silent for a few weeks. When he returned to his typewriter, it was to compose a dispassionate summary of their dealings with Burns. Flood was at his most admirable in this crisis; having devoted some three years to his own attempt to write Wyatt's life, risked his business and his health, endured Josephine's constant criticism, and suffered those insulting letters of rejection, he responded with dignity and self-awareness. Burns was "well provided for with time, pace, and materials, and was without financial worries. Writing a story, surely, is far more than a mere matter of pen and ink, even among experienced writers." A sense of humor glints through Flood's usually dour prose, as he considered how Burns would handle the story of Wyatt Earp without Josephine's interference. "If [Burns] tells [Wyatt's] story in the same manner in which he wrote *The Saga of Billy the Kid*, as he probably will, or has, he will call a saloon a saloon, a church a church, a spade a spade, and a blankety-blank a blankety-blank, and the public will like it."

"Don't be cross with me," he begged Josephine, and signed off, "Your friend."

BURNS RESPONDED TO the Earp tempest with calm confidence. "Mr. Earp," he predicted, "I think you will like my book." By the time Josephine finally read *Tombstone: An Iliad of the Southwest*, she had to admire his portrait of Wyatt Earp as the "Lion of Tombstone," the complete embodiment of the code of the West. Burns had captured Wyatt's handsome natural grace as well as the roots of his character: "He had been roughly moulded by the frontier and he had the frontier's simplicity and strength, its sophistication and resourcefulness, its unillusioned self-sufficiency. He followed his own silent trails with roughshod directness. He was unaffectedly genuine . . . he was incapable of pretense of studied pose . . . he was cold, balanced, and imperturbably calm."

Burns had used his brief interviews with Wyatt to good effect as he drew the most complex portrait yet of Doc Holliday as the "cynical philosopher." As opposed to Flood's painfully contrived dialogue, one could imagine Wyatt Earp actually having said, "Doc was one of the finest, cleanest men in the world, though, of course, he was a little handy with his gun and had to kill a few fellows." Josephine must have been pleased to hear Johnny Behan described as "a bustling, self-important man, never too busy to stop and shake hands or clap somebody on the shoulder with a great show of friendliness."

Josephine's greatest relief came from discovering that Burns had not disturbed the buried secrets of Mattie Blaylock and Josephine Behan. She could find no fault with Burns's closing image of Wyatt's unnamed wife, with whom he was "enjoying his declining years in peace, comfort, and prosperity."

Burns had no reason to be kind. He hardly knew Josephine and had already demonstrated his willingness to trample on her preferences. Although he conducted many interviews in stitching together the story of Tombstone, he completely skipped over the Earp wives and Josephine's role in Tombstone. Either he missed the clues or he found them irrelevant. After all, he was writing a book about the code of the West as dramatized by male frontier heroes. Maria Spence was mentioned once as "Spence's wife," a cameo appearance that was essential because of her role in identifying the men behind Morgan's assassination.

With her anger at Burns subsiding, Josephine hardly knew how to proceed. She indulged in some recriminations at Flood's expense, reminding him that "there many things I DID not like but you were both very stubborn and would not hear of it." There clearly was "a mistake somewhere" in Flood's writing, as Hart said, but the only solution seemed to be a new start with a different writer. Wyatt contacted Frederick Bechdolt, a prolific writer who had interviewed some of the Tombstone old-timers, including Behan's deputy sheriff, Billy Breakenridge, and published novels as well as a popular series of articles in the *Saturday Evening Post* and *Adventure Magazine*. Bechdolt turned Wyatt down.

Josephine had no immediate solution, but her natural optimism reasserted itself: "Don't worry, something will turn up for us."

She did not have long to wait. Despite Flood's concern that Burns's unauthorized biography would corner the market on Wyatt Earp, another book was about to turn up, this one far more dangerous.

ALBERT BEHAN OCCASIONALLY visited Josephine and Wyatt in Los Angeles. Long after his father's death, he remained friendly with both of them, and particularly loving toward Josephine for the brief time that she took care of him as a child. On one of these visits, they reminisced briefly, and then Albert came to the point.

"I think I ought to tell you something," he said. Albert had recently seen his father's former deputy sheriff, Billy Breakenridge, who mentioned that he was now writing a book in which "he's going to burn you and your brothers up—says he is giving you hell." Wyatt looked incredulous, but Albert insisted that he was reporting the conversation accurately. "It made me sick when I heard what he was up to. . . . I thought I should tell you. . . . I think I'd better see Clum about it too."

Alive to the irony of this, Josephine commented: "It seems a bit strange as I think of it that the son of Sheriff Behan should show this interest in the reputation of his father's political enemy. But the character of the two men—Wyatt Earp and the sheriff's son—answers that, and the friendly gesture on the part of the younger man is a compliment to both."

Breakenridge had been at Johnny Behan's side when they were chasing Wyatt. He was a long-standing enemy, but nearly forty years after the Vendetta Ride, he turned up unannounced at the Earps' winter camp near Vidal with a cheery "Hello there, old timer" to Wyatt. Josephine felt dizzy at hearing Billy's "high pitched laugh" and recalled the "bitter heartaches and misery" that Breakenridge and his group had caused them. Despite her mistrust, she served him a big home-cooked breakfast with all of her specialties—fresh biscuits and strawberry jam, eggs, bacon, and coffee—while he and Wyatt reminisced, each of them avoiding any direct reference to the O.K. Corral and its aftermath. Afterward, she and Wyatt agreed that it was high time to set aside old grudges, though Wyatt agreed that Break-enridge would "still bear watching even yet."

Breakenridge was broke but anticipated a big fee from his connection to one of the biggest inheritance lawsuits of the time: the Lotta Crabtree estate. The case hinged on the status of common-law marriages in Tombstone, 1881, a subject that must have given Josephine and Wyatt some pause. Lotta Crabtree had been one of the highest-paid performers of her day, and a savvy investor. Her only sibling, Jack Crabtree, had died before her, and she had left her considerable fortune to charity. Jack Crabtree, however, had lived in Tombstone with a common-law wife and a daughter who survived him. An estimated one hundred claimants had already sued for a piece of Lotta Crab-tree's large estate, but only one had been a baby in Tombstone.

Breakenridge had been recommended by Endicott Peabody to help the estate identify possible witnesses. As he hoped, Wyatt immediately remem-bered the Crabtrees and their baby, and agreed to give a deposition and to help locate other possible witnesses from Tombstone.

Wyatt testified in his understated style. He and his brothers knew Jack Crabtree, and he recalled meeting Annie Leopold Crabtree as "wife of Crabtree" at the Tombstone ice cream parlor on Fourth Street between Allen and Fremont: "I liked ice cream," he said, shrugging. Asked to con-firm the breakdown of law and order in Tombstone, he countered that it was "not half as bad as Los Angeles."

Breakenridge had apparently not briefed the lawyers about Wyatt's own complicated marital history. Wyatt was there as a distinguished Los Angeles businessman giving his opinion on the status of marriage in Tomb-stone. He was not there as a man who had abandoned two common-law wives and was still living with a third.

Wyatt stoutly defended common-law marriage as indistinguishable from marriage sanctioned by state or church:

*Were there or not people in Tombstone going as husband and wife that you didn't know whether they were married or not. You had never seen their marriage license, but you took them as man and wife?*

*YES.*

*Were Jack and Anna Crabtree taken the same way?*

*YES.*

*Were they any different than any others?*

*NO sir, none at all. There were lots of good married people there.*

In the pre-DNA era, nobody could prove Carlotta Crabtree's parentage. The lawyer for the defense argued that common-law marriages were meaningless in Tombstone. Jack Crabtree's daughter received no inheritance.

"I COMMEND TO your consideration Colonel Wm. Breckenridge [sic] a famous peace officer of the old fighting southwest," wrote author William MacLeod Raine to his longtime editor at Houghton Mifflin, a Boston-based publishing company. After a chance encounter with Billy Breakenridge, Raine agreed to read his memoir. It was not quite ready for publication, he warned his editor, but "full of good stuff," and since Breakenridge was one of the only living representatives of the "old time sheriffs who brought law and order to the frontier," Raine thought the public would embrace the book. A successful Western writer himself, Raine was too busy to edit the book, but he offered to write an introduction and recommended Walter Noble Burns or Frederick Bechdolt as coauthors. Raine encouraged Breakenridge to give his book the title *Helldorado*, using a memorable nickname for Tombstone that reportedly originated with a disgruntled miner who complained that he came to Arizona's "El Dorado" to try his fortune but ended up washing dishes in a hellish place. Raine had used the title for a long article in *Liberty* magazine and graciously suggested that it would do well for Breakenridge.

Just one word of caution, Raine warned: "Old Wyatt Earp is still on deck" and might threaten Breakenridge with a libel suit, as he had already attacked Raine for his "Helldorado" article. "He has told his story so long he thinks it is true."

BILLY BREAKENRIDGE WAS living alone at the Old Pueblo Club in Tucson, where he served as the president of the Arizona Pioneers Historical Society. Walter Noble Burns's book had fired him up; it was well written and entertaining, "but as a history it is all buncomb. It is the story told to him by Wyatt Earp while Burns visited him last fall." As for the "Earp Gang," as he called them, Breakenridge scoffed at the idea that they were honest officers; Burns and Frederick Bechdolt had it all wrong. They had fallen under Wyatt Earp's spell. Breakenridge was particularly incensed with Wyatt's claim to have killed the infamous Johnny Ringo during the Vendetta Ride. "A mass of lies," Behan's former deputy declared.

"I am not an educated man," he conceded, and "chasing train robbers was sport alongside of trying to write a book." Still, Breakenridge devoted himself to the task: he researched the original testimony from the Spicer hearing, conducted interviews with old-timers, and ferreted around the Tombstone newspaper archives.

One of Houghton's most experienced editors, Ira Rich Kent, agreed to edit the manuscript, and at eighty-one, Breakenridge was extremely gratified to find himself the published author of *Helldorado: Bringing the Law to the Mesquite.* He was particularly thrilled with compliments from John Clum, who had helped with some fact-checking, and now addressed him as "my dear 'Helldorado' Breakenridge," congratulating him on a "rip-roarin'-snortin' title" and a surefire best seller.

Breakenridge and *Helldorado* replaced Burns and *Tombstone* as the target of Josephine's wrath. Where Burns had idealized Wyatt, Breakenridge cast him as the villain of a lurid tale, portrayed with a sneer and innuendo as the leader of the "so-called law and order party." Never identifying the *Nugget* as the house organ of the cowboys, Breakenridge relied on its highly politicized reporting. Wyatt escaped gunshot only because he wore a bulletproof "steel vest," Breakenridge contended. Behan was treated sympathetically, and of course Breakenridge gave himself a starring role.

Wyatt was genuinely shocked. He wrote to Breakenridge with uncharacteristic pathos: "I have always felt friendly towards you, and I naturally thought you had the same friendly feeling toward me." Despite Albert Behan's advance warning, it was too late for an injunction, so Josephine orchestrated a flurry of letters of protest to publisher Houghton Mifflin, and to their first-respondent group of well-placed friends. A complaint to William MacLeod Raine yielded the cool disclaimer that he had not independently verified statements made in the book, but deferred to Breakenridge.

Burns's book was annoying, but it did not rip open old wounds as *Helldorado* did. Because Breakenridge was a Tombstone eyewitness, his account would carry considerable credibility. "Very interesting," Wyatt noted sarcastically to Breakenridge's assertion that the Clantons and McLaurys were unarmed. "This probably explains how Virgil Earp, Morgan Earp, and Doc Holliday were wounded during the fight." With more nerve than logic, Breakenridge even copyrighted one of Wyatt's most recent photographs.

Josephine's frustrations multiplied. Flood had failed as a writer. Burns had published a flattering version, but the Earps had no share in his success. Breakenridge accused the Earps of heinous acts of violence and corruption. Other accounts were reported to be in the works: even George Parsons was talking about publishing his diaries. Sympathetic to Josephine and Wyatt's distress, Hart suggested other writers such as Rex Beach and Walter Coburn, but neither one was available.

Wyatt's legendary strength was waning. He was in constant pain from a chronic bladder infection. He refused to consult a physician until Josephine delivered an ultimatum. As both of them feared, a physician in Los Angeles recommended immediate surgery. When Josephine returned from making all the hospital arrangements, she found Wyatt packing his suitcases for the desert. "You can have an operation if you want it," he told her with a grin.

They left that afternoon. It would be his last winter in the desert.

"YOU DO THE telling and I, the writing and whipping into shape," proposed an enterprising young writer named Stuart Lake to Wyatt. Just before Christmas, 1927, Lake tracked Wyatt down in the desert to propose that they collaborate on a biography. He had once worked as a press agent for Theodore Roosevelt, and had attended but never graduated from Cornell—credentials that meant nothing to Josephine or Wyatt, until he dropped the one name that did matter: Bat Masterson.

Lake's approach was straightforward and businesslike, promising less a masterpiece of art or history than a work for hire: "I am certain that we could work out a plan including division of labor and remuneration that would be mutually satisfactory." Wyatt responded with some warmth, but postponed their next meeting for a period of time that stretched from a few weeks to six months. He was ailing all that winter, some days entirely unable to leave their camp.

Josephine knew that this was likely to be their last chance. Their fortune had slipped through their fingers. Wyatt was no longer working, and as their money pressures mounted, she found the responsibilities of caring for an aging husband, even one as beloved as Wyatt, to be sometimes overwhelming. Their dream of prospecting bonanzas was long over. They had moved to another Los Angeles bungalow, this one at 4004 West Seventeenth Street, and for the first time, Wyatt acknowledged that his health "was not as rugged" as he would like. They were watching every penny, annoyed by the barrage of books and films about Tombstone and Wyatt Earp, none of which returned any money to them. To supplement Hattie's monthly subsidy, Josephine was forced to borrow money from old friends such as oil tycoon Edward Doheny, who had once worked with Wyatt at the Oriental Saloon and was now one of the richest men in America, living in a grand French Gothic chateau in the same West Adams neighborhood as the Earps' modest bungalow. It was harder to get an audience with Doheny than with the pope, quipped the *Los Angeles Times*, but for Wyatt's sake, Doheny admitted Josephine and paid some of her most pressing bills.

Lake was offering them a fifty/fifty "horse-high, bull-strong, and hog-tight" split. While he had no publisher yet, Lake inspired confidence that he could be the one to write that clean and lively story.

By the early fall of 1928, they'd reached an agreement, and Lake was ready to sit down with Wyatt Earp.

Although Josephine dreamed of securing Wyatt's legacy, she could hardly have imagined that she would succeed in commissioning a work that would endure for generations and become the sturdy cornerstone for Wyatt Earp's reputation into the twenty-first century.

LAKE WAS AWARE of Flood's previous attempt, and requested copies of his detailed notes and drafts. Head held high, Flood turned it all over and said, "I am not interested for one moment as to financial remuneration; the purpose is to square Mr. Earp." Flood's honorable stance was the only compensation he would receive for his patient, if uninspired, work as Wyatt's first authorized biographer. Nor did he receive any credit from Lake, although Lake described "reading and rereading the clippings and Mr. Flood's manuscript" in the first few months as he prepared for his interviews with Wyatt.

His working notes show continual reminders to himself to "see Flood for detail." In later years, Lake would assess Flood's contribution with more ego than honesty and eventually dismissed it entirely: "Flood's so-called manuscript of memoirs contributed exactly nothing to my job: literally, I kept as far away from it as possible. . . . Wyatt never dictated anything of his career to him . . . beyond one quick skimming I never read it until after my book was out."

Wyatt tried again to cooperate but was no more forthcoming than he had been with Flood. Lake found his subject "delightfully laconic, or exasperatingly so." He considered whether Wyatt's long silences were due to the constant barrage of criticism about his past actions. "You have nothing to apologize for," he assured Wyatt, "nothing that will not bear the light in the eyes of any open minded judicious man or woman. . . . You were one man who had nothing to fear from history. . . . Things have changed somewhat it is true, but most of the world knows that in the days with which the most of our story will deal men lived differently than now. We must be as frank about that as about anything else."

However, Wyatt hardly needed Lake's reassurance; he was not a man given to self-doubt, and he had a blunt appreciation of his own strengths and weaknesses. His silence was simply his style. "I was pumping, pumping, pumping, for names and incidents and sidelights; all of which Wyatt could supply but none of which he handed out in any sort of narrative form," Lake complained. To bring color to Wyatt's dry recitation, he eventually resorted to putting words in Wyatt's mouth, as he later admitted: "Wyatt never 'dictated' a word to me. I spent hours and days and weeks with him—and I wish you could see my notes! They consist entirely of the barest facts."

Lake lived in San Diego but came to Los Angeles to interview Earp about six times. Josephine was always there. Lake was restless about her presence and her frequent consultation with lawyer Bill Hunsaker, warning that this kind of "red tape" would impede his progress. There was talk of a visit to Tombstone, but Wyatt was too weak to travel. Josephine responded on Wyatt's behalf to Lake's long lists of questions, correcting facts along the way and debunking myths such as the widespread notion that gunfighters tallied kills with notches carved in their guns. It was Josephine to whom Lake turned most often for photographs and clippings. She threw herself into this new role as research assistant, consulting with her family, with Allie Earp, with Bat Masterson's family. Sadly, much of what Lake sought had been destroyed in the 1906 San Francisco earthquake and fire.

By late fall, with his research well under way and several interviews with Wyatt under his belt, Lake complained playfully that the writing was going too well. He finally read *Helldorado*, useful mostly for its "mistakes and misstatements," but he was concerned about the copyright that Breakenridge had imposed on Wyatt's photograph. A new portrait would have to be taken. In the meantime, Lake made plans to join Josephine and Wyatt in Vidal for the winter, and was ready to pack up his typewriter and books. But when the time came, Wyatt did not have the strength for the move. Then Lake himself fell victim to a serious case of the flu. Bedridden for months, he had no way of knowing that he was losing the last precious months of Wyatt's life.

WYATT'S CIRCLES GREW smaller in the waning months of 1928. Bill Hart wrote often, with warm valedictories about "how glad I am to be considered your friend." A steady stream of friends and family continued to visit the bungalow. Josephine's grandniece Alice remembered her last visit: "I can still see him. He came to the door, straight as an arrow." John Clum was a frequent visitor. Playwright Wilson Mizner came by, entertaining Wyatt with songs from his old repertoire at Considine's Hall in Nome. Tom Mix came from the movie set to spend hours with Wyatt; according to screenwriter Adela Rogers St. John, one of the last journalists to interview Wyatt, he and Tom Mix were serious readers of literature and history, as well as enthusiastic consumers of newspapers and magazines. "A high grade man of the green cloth," the local bookstore clerk noted on Wyatt's file. He and Tom Mix read Shakespeare together, especially *Hamlet,* so talkative a man that "he wouldn't have lasted long in Kansas," joked Earp. Always eager to emphasize that they did not really kill a man before breakfast in Tombstone, Wyatt observed "there are more corpses in *Hamlet* than there was in the O.K. Corral, and with less reason." After all, he added, we killed none of the wrong men like Hamlet did, with a nod of sympathy to "poor old Polonius."

Even in his last year of life, Wyatt had a powerful effect on women. Adela Rogers St. John declared that she would never forget seeing Wyatt as he rose from a chair to greet her, "straight as a pine tree, tall and magnificently built. I knew he was nearing 80, but in spite of his snow-white hair and mustache he did not seem, or look old. His greeting was warm and friendly but I stood still in awe. Somehow, like a mountain or a desert, he reduced you to size."

Nineteen twenty-eight was a season of loss. Tex Rickard and Newton Earp died the same week. Tex Rickard was nearly twenty years younger than Wyatt, while Newton was ten years older. Josephine considered not telling Wyatt, but his doctor advised her to be honest. Tex's untimely death from a ruptured appendix was mourned by thousands of people who filed past a gaudy bronze casket in the middle of Madison Square Garden. Newton had a Grand Army funeral. The brothers had not been particularly close, but the double loss still hit Wyatt hard.

"I can't plan any more to climb the hills and hit the drill," Wyatt mused to Hart. He refused to consult a physician until Josephine recruited Dr. Fred Shurtleff, a prominent physician who was president of the Los Angeles County Academy of Medicine and had served as a deputy sheriff in the frontier and a police captain at Alcatraz. Wearing boots and slinging a saddlebag over his shoulder, Dr. Shurtleff looked every bit the "cowboy doctor" who belonged at the bedside of Wyatt Earp.

On Election Day 1928, Wyatt left the house one last time to have another photograph taken for Lake's book and to cast his vote for Al Smith, the Democratic candidate in the presidential election. Wyatt may have once been a staunch Republican, but the old saloonkeeper had announced that he would cast his vote for whatever candidate stood against Prohibition. Josephine apparently never exercised her right to vote.

Wyatt spent Thanksgiving Day in bed, with his big cat Fluffy nearby. Flood and a few other close friends came to visit; Josephine would recall later how pleased he was to have company on a day that they had always celebrated together, complete with his favorite ice cream dessert.

On January 12, 1929, John Clum visited with a friend from Alaska, and there were some hollow jokes about whether Wyatt might return to Nome. Dr. Shurtleff stayed all day, as did a nurse. During Wyatt's long and restless night, Josephine kept watch. It would always puzzle her what he meant when he sat up suddenly, and said, "Supposing—Supposing . . ."

"I wish I knew what was troubling him." Josephine turned his last words over in her mind. "I should like to finish the sentence for him."

On the morning of January 13, 1929, Dr. Shurtleff was reading aloud from Alfred Henry Lewis's 1905 book *The Sunset Trail*, about the frontier exploits of Bat Masterson and the Earps, and had just finished a passage about Doc Holliday when Wyatt died.

THE PASSING OF Wyatt Earp was a national news story. "Out from the colorful past of the old West stepped these friends of Wyatt Earp," began one article that named each of the remarkable pallbearers, including John Clum, William S. Hart, Tom Mix, Wilson Mizner, Charles Welsh, and Judge Hunsaker. The sight of the distinguished group, many of them leaning on canes and weeping openly, struck a chord of nostalgia as "a reunion of the sturdy men and women who knew Wyatt as a wiry, six foot two gun officer of the law in mining town, cow camp and almost anywhere along the frontier where trouble was apt to pop loose." Although Wyatt was most often eulogized as frontier gunman and courageous defender of law and order, he was also beloved by the Hollywood film community. Dr. Thomas Harper presided over the open casket service, which was held at the Pierce Brothers Mortuary, a chapel of the Wilshire Congregational Church, where Wyatt and Tom Mix had been frequent visitors. Edward Doheny sent an elaborate floral arrangement. Lake was there, still weak from the flu, summoned by a telegram from Josephine.

Josephine did not attend the funeral. In her place, her sister Hattie rode with John Flood in the cortege. Longtime friends Grace Welsh Spolidoro and her mother helped with the details of Wyatt's cremation and served as witnesses, together with Flood and Hattie.

Josephine's absence attracted little notice. This was a day for the public to say good-bye to Wyatt Earp. She may have been overcome by grief or unnerved by the long months of constant nursing. As she knew from previous crises, such as her friend's childbirth in Alaska, Josephine's bravery had its limits. Unlike Wyatt, she did not face the darkest moments of life and death with equanimity.

Josephine considered bringing Wyatt's ashes to Vidal, but her thoughts returned to her childhood and to San Francisco, Wyatt's favorite city. It had been decades since she had any overt Jewish affiliation, but in all the difficulties she was facing, here was one relatively simple, available choice.

She waited a long six months to decide, but finally, accompanied by her nieces, she took the train from Los Angeles with Wyatt's ashes in an urn, held tightly in a satchel on her lap, braced against the rattling and lurches of the train, and brought him to the Hills of Eternity cemetery outside of San Francisco. It was the place where her parents and brother were buried, and there was room for Wyatt and, someday, for Josephine.

Never suspecting that Wyatt Earp was buried in a Jewish cemetery, his acolytes would search for another thirty years before finding his grave.

# 5 | JOSEPHINE'S LAST TRAIL

*L*ESS THAN a year later, some of Wyatt's pallbearers found themselves back at Tombstone.

In 1929 the current editor of the *Tombstone Epitaph* and a group of local boosters suggested creating an event called "Helldorado" as a rousing celebration to mark the old town's fiftieth birthday. Tombstone badly needed some good news; having survived fires, the death of the mining industry, and the shootout at the O.K. Corral, the town continued to lose jobs and prestige to nearby Bisbee, which was now vying to replace it as the seat of Cochise County.

As a move to demonstrate political viability, Helldorado failed. The county seat moved to Bisbee. As a publicity stunt to launch Tombstone's next life as a tourist destination with a whiff of the frontier West, it succeeded beyond its creators' wildest dreams.

No stagecoach or horses required: Tombstone was now an easy drive from Tucson in a newfangled automobile, just a nice day trip along U.S. Highway 80, the "Broadway of America." The plans for the event came together very quickly. Helldorado would be a four-day extravaganza where the whole town spruced itself up to welcome its guests, many of them wearing their best western whiskers and boots, a street party where history took a back seat to braggadocio and pageantry, complete with parades, bands, contests, and brightly painted Indians in war bonnets. "They cleaned the Bird Cage Theatre out, the first time in forty years, and every male within 30 miles is trying to raise a beard," joked Harrison Leussler, who attended the first Tombstone reunion as the western scout for Houghton Mifflin, the Boston publishing company that had carved out a strong market in frontier-themed books. Although Tombstone lost many of its earliest buildings in the fires, Schieffelin Hall and the Bird Cage Theatre were still standing. And of course, despite modifications over the years, there was still an O.K. Corral.

Of the 400 pioneers who were invited, 343 attended. Fox sent a crew to film the entire event for later release in movie theaters. As a highlight, visitors were invited to attend dramatic reenactments of the Earp-Clanton gunfight, with cowboys and lawmen shooting blanks at each other three times a day. No murder, no mayhem, and no booze.

Former mayor John Clum led the parade. He had accepted the organizers' invitation to join an honorary advisory committee, along with Breakenridge, Burns, and others. But Clum came home disgusted by "this style of rip-roaring, Helldorado publicity for poor old Tombstone." Nothing had replaced mining as the engine of economic development, and Clum saw no future for the town other than the exploitation of its violent past.

Hearing Clum's account, Wyatt's friend Fred Dodge was glad he had stayed home. "The battles fought for law and order in Tombstone were no moving picture affairs," he wrote to Clum. "Good men, who were our friends, met wounds and death there. It is an offense to us and to them to reproduce these things as an entertaining spectacle, an incident, for it is not possible to show what necessity lay back of them and made them inevitable."

For the rest of the world, opening day of Helldorado would forever have a different significance. October 24, 1929 would thenceforth be known as "Black Thursday," the day when the American stock market crashed.

Josephine was not invited to Tombstone. She would not have gone anyway, but when she heard about the O.K. Corral reenactments, she started writing letters to friends and government officials. Her lawyer, William Hunsaker, patiently instructed her in the laws of libel, which held that she was not entitled to sue for defamation of a dead person, no matter how beloved or deserving. Arizona governor George Hunt answered her objections with a promise to take up the concerns of Wyatt Earp's widow with the mayor of Tombstone. The mayor of Tombstone followed up with more letters to Josephine from the town's leadership and the organizers of Helldorado, all reassuring her that they were "pro-Earp" and that the staged shootout was "history and reenacted just as it happened." The current U.S. marshal took the opportunity to blame Johnny Behan for everything—and then inquired when he could expect his copy of Stuart Lake's book.

The cowboy faction was no more pleased with Helldorado than Josephine: they had already complained that the reenactment was a one-sided glorification of Wyatt Earp and his murderous brothers.

HELLDORADO AND THE stock market crash were among many sources of distress for Josephine in 1929. Wyatt had been her anchor for nearly fifty years. They had no permanent address, but it had always been the two of them in whatever temporary place they called home, rarely apart for more than a few days. His friends and his enemies had testified to the power of his presence and his strong will; all that had been taken away, and Josephine hardly knew what to think or do. "I miss my dear husband," she lamented in almost every letter she wrote, regardless of the subject or recipient.

She went back and forth between rented rooms in Los Angeles, her sister's home in Oakland, and the desert camp, but no place felt comfortable. Her sister had been her closest friend in the years leading up to Wyatt's death, but now she too was suffering from a combination of business problems, ill health, and anxiety. Hattie's biggest concern was her daughter Edna, who had recently lost her second husband, which left the young mother of two alone again.

Close friends drifted away. The younger members of the Welsh family, with whom Josephine and Wyatt had stayed for long periods of time in the desert and Los Angeles, had a strong preference for Wyatt over Josephine—as if it were necessary to choose between them. They made no secret of their contempt. "She was worthless, I didn't like her, folks just put up with her on account of Mr. Earp," Christenne Welsh said. At a time when money was tight and families were struggling to feed their children, Christenne and her sister Grace believed that Josephine was stealing food from their mother's kitchen. They accused Josephine of leaving a frail Wyatt alone in an empty house with nothing to eat, while she spent her days playing cards, commuting on the "big red" train from Los Angeles to San Bernardino. In their view, Wyatt was a "normal gambler" who enjoyed a friendly poker game in the Vidal country store, while Josephine was a "compulsive gambler" who would return from a day at the tables and head over to the Welsh home to scrounge for food. If she was seen coming up to their door, their mother told them to hide under the bed, hoping that she would go away.

While the Welsh family found it easy to turn away from the widow Earp, John Flood remained steadfast. Drawing on a surprisingly deep well of patience and compassion, he endured Josephine's querulous complaints that he was ignoring her or mismanaging her mining properties. The two of them shared the sharp pain of grief and loss. Flood remained committed to Wyatt's legacy and successfully petitioned the postmaster general in Washington to rename the tiny town near their mining claims as "Earp," recommending the proposed namesake as "an estimable citizen and much beloved by all who know him in this country." He continued to handle Josephine's considerable business correspondence and urged her to sell the mines and, above all, to clear her mind of business conflict: "Yes I know the weather is hot, the flies are a pest, and the mosquitos are worse. . . . Let us set aside the gossip, the envies, the jealousies and all the other useless things of life, and save the situation."

Josephine was particularly anxious about the fate of the Happy Day mines, once the source of such heady expectations. Mine transactions required site assessments and good faith negotiations for which she had neither the training nor the temperament. Wyatt had always handled their mining affairs, with help from John Flood and other experts. Josephine was understandably a novice, but instead of trusting her advisers, in her inexperience she was prone to a suspiciousness that bordered on paranoia. She would agree to a deal, only to back out at the last minute, alienating the professionals and even some of the friends who were trying to help her. She could swing from grief-stricken widow to aggressive, arrogant negotiator in an instant, as she did in response to a friend who offered assistance: "I do wish I could have my dear husband back again, I have surely suffered since he is gone, but I am going to *fight* for what is mine to a finish, and I have good backing too."

As the terrible year of 1929 came to a close, Josephine was a woman alone with her thoughts. She was nearly seventy, and struggling to make decisions without consulting Wyatt, as she had during her entire adult life. Flood did his best to lift her spirits. "Take hold on life all over again; you are just in the middle of it," he exhorted with a burst of religious fervor. "I know that grief is not easy to bear, but life is a riddle—who knows the answer! A wonderful Creator—how grand is the world, all full of His glory!" He urged her to seek distraction in golf, tennis, swimming, horseback riding, taking in the new movie at Grauman's. Although he suggested gently that she consider hiring a private secretary, he volunteered to answer the crush of correspondence after the funeral. Some of those who sent condolence cards and flowers, like Edward Doheny, received individual responses under Josephine's signature. Flood also composed several form letters, one for admiring peace officers, and one for reporters who were asked to respect Wyatt's wish for "silence in the news columns."

To Stuart Lake's dismay, he became the new center of her life. She sent letters and telegrams, called him at home, and asked him to visit frequently. She wanted assurance that he was making progress on the book. Lake appealed to Bill Hart for help in keeping Josephine at arm's length, and also asked for a recommendation to his publisher, whetting Hart's appetite with hints that Wyatt had told him things that no one else knew, not even Josephine. With a mild reproof, Hart reminded Lake that Wyatt's death was still a fresh wound, but then agreed to write a strong personal recommendation to his editor at Houghton Mifflin, urging him to make a deal for "my dear friend's life story."

Acting as his own agent and lawyer, Lake negotiated with considerable skill and prescience. He received no advance, but he retained secondary rights such as serialized magazines and movies, downplaying to the publisher what those might be worth in the future. Josephine's participation was written into the publishing contract, indicating an equal division of royalties. By the fall of 1929, Lake sent Josephine a contract for her signature, with some triumph that the terms of the deal were "exactly as we wished them—maybe a little better than I had full right to expect."

Josephine constantly asked to see Lake's work in progress so that she could correct errors and monitor his adherence to the "nice clean story" of Wyatt as peace officer, loving husband, and frontier leader. Without consulting Lake, she promised the book's dedication to Bill Hart.

She had moved from the Seventeenth Street bungalow she had shared with Wyatt to the nearby home of her Tombstone friend, Sarah Lewellen. Her main activity was research for Lake, "ransacking" her memorabilia and reaching out to the dwindling number of Wyatt's friends from Dodge City and Tombstone. She found photographs of Wyatt and his brothers, but Lake's request for an Earp shotgun proved more difficult. Josephine and Wyatt had stored several boxes and trunks in San Francisco, which were apparently destroyed in the fire of 1906. Bat Masterson's widow had little to offer, since she had given Bat's personal possessions to Alfred Henry Lewis, author of *The Sunset Trail*.

One of Josephine's best sources of information about the Tombstone years was her sister-in-law, Allie Earp. Allie and Hildreth had enjoyed the Sunday visits with Josephine and Wyatt, but now the last of the Earp brothers was gone. When Josephine came around alone, she was received with polite cordiality. Allie's memory was sharp, her tongue even sharper, and she knew far more than Josephine about the Earps in Tombstone. She reminisced fluently and cheerfully. She too had lost most of her memorabilia, including a scrapbook that was destroyed in a hotel fire in Oklahoma, but she still had a few treasures that she passed to Josephine.

"I wonder what Sadie is up to," her niece recalled Allie saying. "She must be asking for all this information for some reason."

Josephine never mentioned a word about a book. Nor did she put Lake and Allie in direct contact; Allie was senile, Josephine told him, an old lady who couldn't tell him a thing.

It hardly mattered, because once Lake had run through the material that he requested from Josephine, he barely spoke to her. When he asked for an extension of their contract to allow him another six months, Josephine reacted with anger and some suspicion. What was taking so long? She turned to Flood for help. Given his own failed efforts to write Wyatt's biography and a growing mistrust of Stuart Lake, Flood must have found his situation acutely uncomfortable. Only a few years before, he had been the one under Josephine's thumb, enduring her demands for speed and her editorial tyranny. "Writing a story is not an easy task; much more time is required, usually, than one may imagine," Flood wrote tactfully, apparently resisting the natural impulse to remind Josephine that he knew her "hints" all too well.

When Lake found himself on the receiving end of Josephine's constant reminders that his book "must be entirely different from all others," nothing in the "blood and thunder" mode, he displayed none of Flood's restraint. Lake had no particular affection or respect for Josephine, who was a constant drag on his time and an annoyance to an experienced, self-confident writer.

Researching Wyatt's early years as a lawman, Lake ran up a considerable tab and billed Josephine for his expenses. He and his wife took an extended trip to Tombstone, soaked up the local atmosphere, and met with some of the people recommended by Josephine and others. "I bought me a pair of overalls," he congratulated himself. In the courthouse vault, he found juicy details on the checkered past of Johnny Behan, who had been indicted several times after Wyatt Earp left town, with charges ranging from "minor thefts of official funds to felonies of scandalous proportions." Lake also discovered that Johnny had run away with "his best friend's wife"—presumably Emma Dunbar. None of the previous Tombstone historians had done anything like this level of sleuthing, he bragged: not Burns, nor Bechdolt, nor Raine, nor Breakenridge.

"I have not tried to interpret Wyatt Earp," Lake pledged. "I have told of the old West as simply as I might." But by the time he began to write, Wyatt was dead, and there was no check on Lake's exuberant press-agent style of history. He was after the legend, not the life, and had no hesitation in making up quotes and pumping up thin anecdotes with additional material that made for a better story, while claiming that all his information was carefully documented and out of the mouth of Wyatt Earp.

Meanwhile, to the dismay of his editors and Josephine, Lake missed deadline after deadline, sometimes taking off a few weeks from the biography to "turn a quick dollar" and churn out some lucrative magazine articles. As the delays stretched out from weeks to months, and Josephine had not seen a single page, she began to suspect that Lake was actually working on a different book. She pounced on the publication of a novel called *Saint Johnson* as confirmation: the tale of a brother's vengeance in corrupt Tombstone was clearly based on the Earps. Despite Lake's strenuous denials, she accused him of using William R. Burnett as a pen name.

Aligning Wyatt's life and career to the mores of Prohibition and the Depression would have been difficult for any author, let alone one with Lake's mythmaking talents. In the early days of their collaboration, Lake sought to placate Josephine, sometimes appealing to her friends for advice and moral support. "I have nothing but respect for Mrs. Earp's sentiments, the fullest regard for her idolatry of her husband's integrity," he protested to Hart. Now he hinted to Hart and others that Wyatt had preapproved his manuscript, implying secret knowledge that Wyatt had not shared with his wife: "I have never had the heart to tell Mrs. Earp that Wyatt and I managed to sneak in quite a few talks together, away from her." Fearing that her long-dreaded exposure was at hand, she appealed to Hart, who rebuked Lake with a sharp reminder that "the memory of her husband is very dear and vital to her."

The thought of Lake walking around Tombstone and talking to the old-timers must have struck terror into Josephine's heart. She had every reason to fear that he would stumble over her relationship to Johnny Behan and the sordid tale of Mattie Blaylock Earp.

This time, her fears were well founded. Lake smelled a big story in Josephine's past. He learned that Sadie Mansfield, the prostitute who had been cited as a central figure in Johnny Behan's divorce, had followed Behan to Tombstone, where she raised her prices and became known as "Forty Dollar Sadie." Perhaps Sadie Mansfield was Sadie Marcus, he theorized. Both women were from Europe, and their families had lived in New York before traveling west. That Sadie Mansfield may have been from Germany, and Sadie Marcus from a mostly Jewish province in Prussia, was hardly in Lake's calculus. He shared his theory with Leussler, who had spent decades scouting old Western stories and agreed to use his own sources in Tombstone to probe into the possibility that Josephine had been a prostitute.

"Do you remember in our conversation, the last time we met there was a certain person mentioned called Forty Dollar Sadie?" Leussler wrote. "I just wanted to tip you off to the fact that the wrong supposition was taken when we suspected a friend of mutual acquaintance of being this person, because it is not so. They are two separate distinct people. B.B. [presumably Billy Breakenridge] knew them both and when I next see you I will tell you a damn good story about forty dollars. But don't, for goodness sakes, think it is the person we both thought it was." In case Lake did not get his broad hint that "the person" was Josephine, Leussler continued with a strong recommendation that Lake let "her" see the manuscript as soon as possible.

So maybe Josephine wasn't *that* Sadie, but something still seemed fishy to Lake. Knowing nothing about her brief history with Pauline Markham's Pinafore troupe, but having some evidence to tie her to Behan and believing that she had been a performer, he concluded (and attributed to Bat Masterson) that Josephine had been a dance-hall girl, "the belle of the honkytonks, the prettiest dame in three hundred of her kind." And from those near-facts, he spun a tale close to the truth: "Johnny Behan was a notorious 'chaser' and a free spender making lots of money. He persuaded the beautiful Sadie to leave the honkytonk and set up as his 'girl,' after which she was known as Sadie Behan."

Lake was off in a few ways: the love triangle of Behan-Josephine-Wyatt was not "Tombstone town talk," as he asserted. The inquest and other legal proceedings around the gunfight never once mentioned Josephine. Her relationship with Behan never appeared in the newspapers, nor did Allie Earp acknowledge that Josephine and Wyatt had an affair in Tombstone. In focusing on Josephine, Lake missed the bigger story that was hiding in plain sight: Wyatt Earp, the subject of his biography, had a wife in Tombstone, and that wife had come to a very bad end. Despite Lake's self-congratulations on his extensive research, he failed to find Mattie in his interviews with Wyatt's friends, and he missed the paper trail in the public documents. She was identified as Mrs. Wyatt Earp in the 1880 census, in real estate records, in mining claims, and in the Tombstone newspapers.

"Should I or should I not leave that key unturned?" Lake demanded of

his editor, Ira Rich Kent. Already impatient with Lake's slow pace and pur-
ple prose, Kent brushed off Lake's circumstantial case against Josephine:
"I don't see how you can very well tear away Mrs. Earp's entire covering
and leave her completely exposed after 50 years of what we will assume to
be a faithful and virtuous life." Besides, Kent "already knew the general
facts about Mrs. Earp's status at Tombstone," presumably from his agent
Leussler. So Kent urged Lake to leave out Josephine by name but to include
some "general statement to the effect that it was understood that a part of
at least the trouble between Behan and Earp arose from their disagreement
over a girl"—and get on with the important business of finishing the book.

Never one to admit defeat, Lake did a quick reverse and assured Kent
that he would find another way to get the "woman interest into the Tombstone chapters."

Had Lake discovered the truth about Mattie Earp, he would probably
have buried the story anyway. Would the public cheer for a hero so flawed
that his actions led his abandoned wife to prostitution, drug addiction, and
suicide? Great storyteller that he was, Lake would likely have shrugged his
shoulders and cheerfully relinquished the juicy scandal, since he saw more
profit in building up Wyatt than in using his women like bullets to knock
him down.

Lake and Leussler conspired to keep Josephine in the dark, especially
about the proposed title, *Wyatt Earp, Gunfighter*, which Leussler predicted
will "hit her right between the eyes. . . . My suggestion is that you make a
change in the title, for her eyes, and when the book is submitted if the publisher changes it back as you originally titled the story—why of course you
won't know anything about the matter as the publisher sometimes makes
such a change. Get the point?"

Whenever Lake and Josephine did talk, they quarreled. Lake chose
a photograph of Wyatt in shirtsleeves; Josephine objected that a coatless
Wyatt would be taken for a "tough." She warned that Lake was making
her sick; in fact, she was suffering from severe headaches that had been

diagnosed as some kind of "growth on the brain." Resisting the entreaties of her sister and niece to go to the hospital, she declared melodramatically that she wished to have completed her review of the manuscript before "the unexpected might happen and I should pass away as Mr. Earp." Perhaps she did have a brain tumor, Lake told his friends, but his diagnosis was a lot simpler: Josephine was crazy. "She is undependable, mentally; unbalanced, psychopathically suspicious."

Still convinced that Lake had written *Saint Johnson*, Josephine was furious to learn that the objectionable book was being filmed by Universal. She spent $700 with a private detective to prove Lake's guilt (the detective found nothing). As a last resort, she set off across the country to plead her case directly with Lake's publisher.

This was only her second trip to the East Coast. Before she left, she had dinner with John Clum and then stopped off in Texas to visit with Fred Dodge. They had already heard from Lake, who began courting them before Wyatt's death. With bland hypocrisy, Lake assured Wyatt's old friends that Josephine was still his trusted partner and collaborator: "Mrs. Earp is to share [the book's profits] with me, now that Wyatt is no longer here. That was provided for in our agreement, but would have been the case anyway. We signed the paper as a matter of record for possible publishers. Between us, we needed none."

Clum and Dodge suspected no duplicity on Lake's part and conferred on how they might best serve as Josephine's advocate. "It was sad to see her so worn and troubled, and so alone," Dodge sympathized. "Evidently she is inexperienced in handling business matters. I suppose Wyatt attended to everything." He suggested that she give Clum power of attorney to represent her in the publication business. They shared her disdain for *Saint Johnson*. Those books were worthless, Dodge and Clum agreed, except as "samples of the trash that is being gotten out and sold on the wave of popular interest in the old west. It would make you, and me, and Wyatt, if he were still here, think that we had never been in Tombstone."

When she arrived in Boston, Josephine went straight to Houghton Mifflin's headquarters. Despite Lake's warning that she was unhinged, his editor Kent found himself deeply moved by his encounter with the older woman described as such a harridan by Lake: "She sat at my desk for a better part of an hour, tears rolling down her cheeks in her emotion. She told me of various disagreements with you and with her refusal to meet your requests in some particulars." He found no signs of mental instability, attributing her emotion to the recent loss of her husband of forty-eight years. Reminding Lake that his relationship with Josephine was not really Houghton Mifflin's problem, Kent recommended that Lake proceed with greater "tact, patience, kindliness, and forbearance"—all characteristics that Lake had in very short supply.

Lake finally produced a manuscript. To his surprise, Kent rejected his draft as "bilious and jaundiced," too long and too pretentious. Complaining that it read more like a eulogy than a vivid biography, Lake's editor appealed to George Parsons, who agreed that it dragged along and was overly partisan. This was hardly the warm first reading that Lake had expected.

Josephine returned from Boston, comforted and emboldened. In constant contact with Flood and with her lawyer, she barraged Lake with demands. When he finally forwarded the manuscript, now revised and shortened, she protested that it was still the "blood and thunder" book she most detested, and fired off a long list of complaints. Setting aside any attempt at diplomacy, Lake took direct aim at Josephine's "constant heckling and interference" and shot back: "I have been patient, tolerant, sympathetic, and understanding. I have wasted hours of time trying to explain and re-explain things which have been misconstrued and distorted. . . . The patience that I own is being strained close to the breaking point." He abandoned any pretense of hiding his contempt for what he saw as ignorant meddling, suspecting that John Flood was the instigator. "I don't care what your well-meaning but entirely mis-informed friends and advisors tell you," he retorted to her questions about possible sales to magazines and movies.

In one crucial area, however, Josephine was vastly relieved. Lake's Wyatt Earp was sober and single. Other than one oblique hint, the love triangle was left undisturbed. There are no Mrs. Earps at all in his version of Tombstone, no women or wives for Wyatt, Virgil, Morgan, or James. Only Doc's lover Big Nose Kate and Addie Borland, the milliner who testified at the Spicer hearing, are mentioned. "Mrs. Wyatt Earp" makes one cameo appearance at the very end of the book when Lake inserts her into a scene of affluent domestic bliss: "In San Francisco," he notes (tongue firmly in cheek), "Wyatt had married Josephine Sarah Marcus, daughter of a pioneer merchant in that city who, as Wyatt put it, 'was a better prospector and camper than I ever hoped to be.' Mr. and Mrs. Earp divided their time between their highly profitable oil and mining properties and their homes in Oakland and Los Angeles."

Josephine and Wyatt had faced down deadly opponents, escaped bullets and scandals. They outlived most of their enemies. Together they could have withstood anything, but alone, Josephine had lost her center of gravity. The scaredy-cat emerged at last. Stuart Lake proved too much for her.


JOSEPHINE AND LAKE'S editor surrendered at almost exactly the same moment. Faced with Lake's intractable opposition to any material changes, and his thinly veiled threats of a lawsuit, Josephine was defeated and contrite. From her lawyer's office, she dictated an obsequious letter to Lake, in which she promised to cooperate in the future and to rely on Lake to protect her interests. She gave up all of her demands, including changing the title and restoring the dedication to William S. Hart.

From Boston, Kent accepted Lake's edited manuscript and sent the author a flattering telegram: EARP IS GOOD JOB MAKES FASCINATING READING STOP.

Once Josephine and Kent capitulated, Lake turned his attention from writing to marketing, relying on his experience as a press agent. Early serialization of books was an excellent form of prepublication promotion and Lake sold serial rights with great gusto, negotiating to cover the same travel and research expenses that he already tried to charge to Josephine and to Houghton Mifflin. America's favorite magazine, the *Saturday Evening Post*, published some of his best chapters under the title "The Man who Brought Law to The Frontier." If Josephine had not been so angry at Lake, she would have appreciated the placement, since her nieces used to joke that the Wyatt Earp they knew was more likely to hold the *Post* than a gun.

Houghton Mifflin had high hopes for *Frontier Marshal*. The publicity and advertising campaign stressed that readers were finally going to get the one true story: "Wyatt finally did talk! He spent the closing months of his life telling Stuart Lake the story of the Old West that no one else knew." A reproduction of the handsome jacket illustration was distributed to bookstores, with the caption "He outshot the bad men of the old Southwest." Lake fired off one marketing salvo after another, and Houghton Mifflin complied with most of them. No detail was too small to escape his attention: when he saw a jigsaw puzzle that had been made from the cover illustration, he objected to the fact that the title did not appear, "not a damn word on the box or puzzle" to cross-promote the book. He constantly monitored sales and bookstore inventory, complaining like any other author when friends wrote to say that their favorite bookstore was out of stock. His most creative idea was displaying frontier memorabilia in major office buildings and bank chains along the Pacific Coast, using enlarged photos, shotguns, saddles, and poker games. The window displays were so popular that they caused traffic jams.

*Frontier Marshal* was a runaway success. Wyatt Earp was a child of America's West, now reborn into the Depression and Prohibition, an era of audacious bank robbers and gang wars. America needed a hero, declared Florence Finch Kelley, who reviewed *Frontier Marshal* for the *New York Times*. She compared Tombstone to the fearful lawlessness of Chicago and New York. "What both these cities need seems to be a few Wyatt Earps," she wrote. "A hundred times the best book written on the Western era," gushed the *Dodge City Globe*. "In a day when cool, courageous, steel-nerved men were the rule, Wyatt Earp was the coolest, the bravest, the nerviest of them all," wrote Joseph Henry Jackson in the *San Francisco Chronicle*, while in the *New York Herald Tribune*, Leslie Gannett promised that readers would find that *Frontier Marshal* had "all the exciting qualities of a dime novel, the added value of authentic history, and the curious virtue that it might be used as a Sunday School text of a Hollywood scenario . . . the blood and thunder and stirring human qualities of the great Norse sagas and of Homer, yet is of our own day and generation."

Wyatt's friends gave Lake their own rave reviews. Congratulating himself and Fred Dodge for having stayed alive long enough to see the book's completion, John Clum wrote, "I think it is the best story of the old West that has been published. I am sure that Wyatt would like it." It was indeed "Wyatt's book," agreed Fred Dodge.

A minority of negative reviews complained about the book's length, and the tedium of reading about a cartoonish Marshal Earp wearing a "halo and a robe." A few claimed that the Tombstone episodes were drained of drama, more like a "brief for the defense." A more substantive criticism came from Billy Breakenridge's champion, William MacLeod Raine. Lake was no historian, and his "incorrect and absurd" account of the gunfight deliberately concealed that the cowboys were leaving town when the "gamblers came down to kill them." He must have been hypnotized by Wyatt, Raine speculated.

The popularity of *Frontier Marshal* pacified Josephine immediately, as did the handsome royalty checks that began arriving from Boston, split evenly between her and Lake. Given the dark days of the Depression and the relatively recent publication of western books by Burns and Breakenridge, the sales were remarkably robust. Although Josephine was miffed that Lake sent her one measly complimentary copy, she was pleased with the money, and gratified by the fulsome praise from Wyatt's friends. Lake presented her with copies of some fifty favorable reviews.

However, after all the fine words with which Lake began their partnership, and the contentious negotiations over contracts and editorial direction, he was perhaps hardest on Josephine after the book was published. To Kent he wrote: "She has done nothing but make trouble ever since Wyatt died, for me. She has not contributed one word of help or information, but has devoted herself to spreading scandal about me. . . . Suppose you think I'm hard, but I know Mrs. Earp from Wyatt's intimates, over a long period of years. Long ago, one of Wyatt's best friends told me how to handle her. I hadn't the heart to take his advice, until I became desperate."

Lake and Josephine were like divorced parents with joint custody of Wyatt Earp: the love was long gone, but they could never be free of each other. Perhaps in recognition of that reality, their relationship eventually stabilized. Lake hogged the limelight, but Josephine preferred it that way. Lake had actually managed the writing and publication of Wyatt's biography with considerable skill, she realized, and the weight of positive reviews was convincing. Perhaps too his confrontational manner—with her, with his publisher, and with anyone else who threatened to get in his way—made it absolutely clear that she had no leverage. Even her lawyer advised her to accept her situation, which after all was not so bad, with a steady stream of royalties and the prospect of lucrative movie deals. With the death of George Parsons, and then William Hunsaker, and then John Clum, Josephine was losing her circle of allies.

This did not mean that Josephine stayed out of Lake's way. She tried to stop production of the movies *Law and Order* and *Frontier Marshal* in the belief that they placed too much emphasis on Wyatt as a gunslinger. Although the studios did not need her permission to make a movie about Wyatt, they preferred to avoid any legal entanglements. They sanitized the script for the film version of *Frontier Marshal*, but also downgraded the production. Lake, of course, was furious, because he was counting on box office success to stimulate sales of the book. Josephine mocked the eventual film *Frontier Marshal* as "no more the story of Mr. Earp's life than an animated cartoon of Mickey Mouse."

The news that Lake's wife was pregnant took the last heat out of Josephine's fury. Baby inquiries replaced questions about book sales, copyrights, movie rights, and photographs; she tracked the baby's due date and reminded Lake that he must let her know immediately whether the stork delivered a girl or a boy. She was one of the first to send a gift and a card to baby Marion Carolyn Lake.

When Lake first begged Josephine to give him Wyatt's set of Bancroft's *History of the United States, from the Discovery of the American Continent*, she refused, but later sent the books as a sign of the growing goodwill between them. Besides, she owed Lake money, and hoped he would enjoy owning Wyatt's volumes and "apply [them] on the little debt I owe you." She also began to refer publicly to Lake's book as accurate, interesting, and the last word on her husband's life. To reporters who approached her with questions about Dodge City and Tombstone, she developed a standard response: "In the writing of *Wyatt Earp, Frontier Marshal*, the author, Mr. Stuart N. Lake of San Diego, made a thorough research of the incidents of Mr. Earp's life." In fact, she stated, the book was "dictated to Mr. Lake by my husband."

Lake would never have another success like *Frontier Marshal*. For the rest of his life, he would remain fiercely litigious about anything to do with Wyatt Earp. His belief that he had exclusive rights would eventually bring him into conflict with anyone who felt that they too had something to say about Marshal Earp and Tombstone.

STILL A STRONG woman in her seventies, Josephine tried to resume the nomadic cycle that had once seemed so natural during her five decades with Wyatt. The mining camp at Vidal, which had once been filled with the sound of Wyatt's deep voice, was silent. She rarely slept there, preferring the Grand View Hotel in Parker, where she reserved the bedroom at the top of the stairs, one with a chamber pot, to avoid a long walk in the middle of the night. With her share of the royalties from Lake's book and Hattie's oil wells, she could support herself. In Los Angeles, she occasionally rented rooms in the area where she and Wyatt had lived, but more often saved money by staying with friends like Sarah Lewellen, whom she had known when they were both young women in Tombstone.

She and Flood corresponded regularly, and he typed her business letters as well as the occasional response to a fan or a persistent newspaper reporter. On Thanksgiving 1931 she wrote to Flood from Oakland, where she was staying with her niece, Edna. Josephine was distracted and overcome by memories of a previous November: "Everything was just lovely, but I missed my dear one. I just kept thinking of our nice dinner you and a few others had just three years ago Thanksgiving Day at our little home in Los Angeles when my dear husband was ill in bed but smiling and said you all enjoy yourselves and I will have mine in bed and he was so happy to have you all with him. I can just hear him say Flood I will have my ice cream now. Oh how I do wish they would let him rest in peace."

Now she faced her first Thanksgiving without her sister, who died on April 7, 1936. After Wyatt's death, the two women had become even closer. Hattie sold Lehnhardt Candies and closed the big house in Oakland, moving into a smaller residence where Josephine continued to be a frequent visitor until Hattie's death. "I am very lonely for my dear sister and it seems like I have nothing now to live for, my darling husband left me and now my best dear friend has gone away too," she cried to Stuart Lake.

The loss of her sister precipitated a medical crisis, just as Wyatt's death had. Josephine suffered an acute attack of arthritis in her neck and upper arm, which nearly immobilized her for months. She was contending with money problems as well; now that she was not staying frequently with Hattie, she had more of her own expenses to cover. Josephine wrote to Lake of her desire to sell a bearskin rug that had been given to Wyatt by Bill Hart

and also confided "I am having some trouble about my interest in the royalty from the oil and I will have to wait until it is settled which I hope will
be soon."

Hattie's ashes were scattered among the wildflowers of the Berkeley
Hills, as she had instructed. Her chinchilla-trimmed broadtail coat and Russian sable scarf went to Edna, who received half of her mother's estate, with
the other half left to Hattie's son Emil Jr. If both of her children died, the
estate was to be split equally between their children and "my beloved sister,
Josephine Earp." However, the will was silent on the subject of the wells that
had been legally transferred by Wyatt to Hattie and then leased in 1925 to
Getty Oil. After all the time that Josephine and Wyatt had spent prospecting, it was only these investments that had proved to be significant. They
were now generating enough money to be a major part of Hattie's estate.

Emerging as the strongest voice of the Marcus family's next generation,
Edna believed that she and her brother owed nothing to Josephine. She was
the sole supporter of her two children and had serious aspirations as an artist. If she didn't have to share with her aunt, Edna could live comfortably
on her inheritance. Ironically, Josephine may have seemed too much of a
Marcus to Edna, who was eager to begin her life anew with no Jewish connections to slow her down.

The payments to Josephine, which had once arrived so regularly on the
first of the month, simply ceased.

Josephine inquired, politely at first, and then complained with increas-
ing urgency. After months of entreaties, she filed a lawsuit in 1936 to enjoin
Getty Oil from paying royalties to Hattie's heirs. Neither side really had
their heart in the lawsuit, however, and court dates were delayed so many
times that the judge threatened to drop the case from the calendar. The
relations between Josephine and her family remained cordial during most
of the lawsuit. They would appear on opposite sides of the court during the
day, and go out to dinner in the evening. The plaintiff, Josephine, baby-
sat occasionally for her great grand-nephew, who was the grandson of the

defendant, Edna. In the will that Josephine composed during the lawsuit, she divided her estate equally among her nieces and nephews, referring to each of them warmly by name, leaving a blank for the person or people who would be the ultimate beneficiary of the disputed oil properties, and also leaving another blank for the recipients of future *Frontier Marshal* royalties.

Edna idolized Wyatt, and had fond memories of her many visits to the camp near Vidal. But her feelings toward Josephine were strained by the lawsuit. When the court finally agreed with Edna that she and her brother had sole control of the oil leases, with no legal obligation to share with their aunt, there was little reason for them to remain in contact.

JOSEPHINE SHOULD HAVE been the matriarch of the family, but she presided over an empty table. Hattie's death and the lawsuit permanently loosened the ties between Josephine and her family. While not absolutely penniless or homeless, she was a lonely old woman, worried about the future, and dismayed about the rift with her niece. She was lost without Wyatt and Hattie, and depressed by the reality that her life of adventure had slipped forever from her grasp. The entire country's economic depression and the ominous gathering of war clouds over Europe only magnified her diminished personal circumstances.

But something always turned up for Josephine. She would live to write Wyatt into one more chapter of the American frontier.

IN 1931 THE explorer Lincoln Ellsworth was planning a historic expedition to Antarctica, hoping to complete the first transcontinental flight. An enthusiastic reader of *Frontier Marshal* and other tales of Tombstone, Ellsworth wanted to give copies of Lake's book to his entire crew and to create a shrine to his boyhood hero in the captain's cabin. Best of all, Ellsworth planned to name his ship the *Wyatt Earp*.

Ellsworth wrote to Lake, requesting a photograph of Wyatt in later life. "Charmed to oblige," Lake responded and crowed to his editor, "Can you beat Lincoln Ellsworth's naming his ship?" Ellsworth also contacted Josephine, who realized immediately that anyone who did not know Wyatt's name before would certainly know it now, and would associate him with courage, integrity, and patriotism. With help from John Flood, Josephine gathered together a precious store of memorabilia, including Wyatt's Colt 41 pistol (oddly, the weapon he brought to the Sharkey-Fitzsimmons fight), one of Wyatt's shotguns, and his last pair of wire eyeglasses, worn just one week before he passed away. "I would not part with them if it were for someone besides you," Josephine wrote.

Born to a wealthy family in Chicago, Lincoln Ellsworth had studied engineering and trained as an aviator, successfully flying over the North Pole in 1926. He completed four expeditions to Antarctica between 1933 and 1939, using as his base the Norwegian herring boat *Fanefjord*. This was the boat that he planned to refurbish for his next Antarctic journey. It would carry an entire engine in spare parts below deck, as well as Ellsworth's plane and a wild pig being fattened for a New Year's feast. And this was the boat that would be reborn as the *Wyatt Earp*.

Why change the name? Ellsworth asked, answering his own question in one of his early dispatches. The choice of Wyatt Earp connected his voyage of exploration specifically with the American frontier and American individualism, "all the best qualities in pioneering and development." It had been America's manifest destiny to expand from the Atlantic to the Pacific, and now it was Ellsworth's dream to plant the flag on the southernmost margin of the planet.

Because of public fascination with Ellsworth's exploits, Wyatt's name was featured prominently in the newspapers at least once a month for six years, sometimes in several articles on a single day. In the *New York Times* alone, 227 articles appeared, with headlines such as "*Wyatt Earp* Sails On in Foam-Crested Sea." The name of Wyatt Earp was indelibly linked to a national adventure that captured the American imagination.

The four voyages of the *Wyatt Earp* make for breathless reading even today. The "staunch little ship" fought its way through subpolar storms and vast ice packs—not unlike Josephine's experience during her Alaskan journey. The suspense hardly ended when Ellsworth finally planted the American flag on Antarctica. During his flight across the continent, he was forced down by a blizzard and missing for almost two months. As his crew struggled to reach him, Josephine and the world read eagerly: "*Wyatt Earp* Off on Ellsworth Hunt" . . . "*Wyatt Earp* Braves Dangerous Waters" . . . "Ice Halts *Wyatt Earp*" . . . "Storm Aids *Wyatt Earp*." Ellsworth's rescue dominated the news, even on a day when Edward VIII ascended to the throne of England and Dr. Joseph Goebbels addressed 18,000 cheering members of the National Socialist Party in the new Deutschland Hall, demanding new colonies to fuel the growth of Germany.

Josephine was overwhelmed with pride, as if Wyatt himself were animating the intrepid explorers. The global triumph of the *Wyatt Earp* went a long way toward assuaging the long years of public excoriation—perhaps even better than the praise and the royalty checks from *Frontier Marshal*. Surely she would triumph yet over her petty money problems and private terrors of discovery and humiliation.

Perhaps, Josephine reasoned, the time had come for her to return to the stage.

THE TOMBSTONE MEMOIRS and sensationalized "true life histories" that began with Bat Masterson and continued with the early dime novels had never stopped, and now a new crop appeared. *Frontier Marshal* was the catalyst for much of this activity; Lake's idealized account infuriated people like Doc Holliday's former lover, Big Nose Kate, who was none too pleased with her role as Doc's whore. Kate wanted to set the record straight and was collaborating with a writer. John Clum wrote *It All Happened in Tomb-*

*stone* and published one of the few biographies of a western woman, Nellie Cashman. He became a popular lecturer on the frontier experience. George Parsons died without publishing his diaries, but Fred Dodge's notes for his memoir eventually appeared as *Under Cover for Wells Fargo*.

Josephine was most incensed about *Pioneer Days in Arizona* by Dr. Frank Lockwood, which portrayed Wyatt and his brothers as cold-blooded killers who stalked the Clantons and McLaurys to their deaths at the O.K. Corral. As an academic, Lockwood should have been more impartial and historical, she scolded, more like Lake. Instead, Lockwood relied on Breakenridge's *Helldorado* and produced a book that "teems with misinformation."

It was Wyatt's way to ignore criticism, but incited by Burns, Breakenridge, Lockwood, and emboldened by Ellsworth and Lake's success, Josephine had had enough: "It is time that I had something to say," she declared, "I am writing the story of my life."

She discussed her plan with John Clum and with John Flood, who offered their assistance; like Wyatt, she would need a writing partner.


IN 1936, JOSEPHINE noticed an obituary entitled "Wife of Gunman's Kin Dies," which identified the recently deceased Alice Earp as the wife of Wyatt's cousin and a resident of Los Angeles. Josephine tracked down the family mentioned in the article; they had not known that Wyatt's widow was alive but invited her to visit.

Josephine arrived the next day. Instead of the rotund, windblown Josephine who presided over the Earp campsite, a newly slender and attractive older woman descended from a taxi, carrying a bakery box with a fresh cake inside. She was barely over five feet tall, and instead of her old shapeless housedress, she wore dignified and expensive-looking clothes. This first visit was lively and affectionate, as if they were indeed discovering long-lost relatives. They were eager to hear about her life with their famous cousin. And how exciting for Josephine to discover that these nice Earps were also writers and history buffs.

Perhaps they would be proper stewards of her story.

Perhaps, she suggested, they might collaborate on her memoir.

Her hosts were Mabel Earp Cason and Vinnolia Earp Ackerman, the daughters of William Harrison Earp, Wyatt's cousin. As children in rural Texas, New Mexico, and Mesa, Arizona, they had grown up surrounded by recollections of the frontier and the Earp brothers. Seeing Annie Oakley in Buffalo Bill's Wild West Show was a highlight of their childhood. At a time when laborers were paid a dollar a day, they could hardly believe that Chief Sitting Bull was paid $1.50 for one signed autograph. They never met Wyatt, and to them, Tombstone was important not as the site of the Gunfight at the O.K. Corral but as the place where their parents had chosen to be married, simply because the location was convenient and fun.

They were raised with a spirit of tolerance and independence, and an emphasis on education. A poor but sociable young girl, Mabel recalled walking through "chippy town" to get to school, and befriending some of the prostitutes, who were nice to her and to her hardworking mother. The family went through the worst of the Depression in a Texas farmhouse, where men would come to the back door every day asking for food. The sisters had clothes but sometimes no shoes.

They moved to California in 1934, and Mabel became an art teacher and writer/illustrator of children's books, while Vinnolia was eventually promoted to a senior position as head technical writer for a pharmaceutical company. They were both married; Mabel and Ernest Cason had five children, Vinnolia and Harold Ackerman one son.

The sisters had never written a book together, but responded enthusiastically to Josephine's proposed collaboration. They liked her story, and they liked her. Their husbands were less enamored of Josephine, but both felt that the project had some potential to make money.

Recognizing that Josephine was lonely and living in a state of genteel poverty despite her good clothes, both families invited her to stay with them while they planned the book. Except for the husbands, all of them grew fond of her. What began as a temporary visit lengthened into a cycle in which she would stay with Mabel, then Vinnolia, for months at a time, and then return to other friends in Los Angeles while Mabel and Vinnolia wrote up their notes and continued their research. In the Depression era, taking in elderly relatives and stranded friends was commonplace; the Casons lived in a four-bedroom house, with grown children who were coming and going. Now it was "Aunt Josie" who became part of the extended Cason-Ackerman family.

Josephine brought Mabel and Vinnolia a large stack of research notes that had been started by the friends she had already consulted about her memoir, a short list that included Clum, Parsons, and Hart. And then the three women set to work: the interviews were conducted mostly by Mabel, with Vinnolia taking notes in her expert shorthand. Sometimes Mabel's daughter Jeanne sat in on the interviews, listening with fascination to the stories of the old West. The sisters decided to break the story into two sections: Vinnolia would write about the events leading up to and including Tombstone, while Mabel would work on life after the O.K. Corral.

The sisters were captivated by Josephine's deep love and respect for Wyatt. She cast herself as Wyatt's traveling partner and foil, never as the center of attention. Although Mabel and Vinnolia were both working mothers and respected partners in their marriages, they accepted her willing subservience as charmingly anachronistic. "Wyatt spoiled her, and she was his little pet," Mabel's daughter Jeanne recalled. Josephine was totally devoted to Wyatt's memory, even "worshipful," as well as protective of his image and reputation. A romantic teenager herself, Jeanne felt that she was hearing from a woman who adored her husband, and was sharing the stories of a successful marriage, albeit one never blessed by any church.

But in their private discussions, the adoring wife acknowledged some deep dents in her hero's armor. Wyatt had affairs while they were together,

Josephine confided to the sisters, and she even hinted at the existence of a previous wife. "She told us other things that indicated that, while Wyatt was a fine peace officer and a very brave man, he didn't have the best of principles where women were concerned," Mabel recalled. Josephine admitted that she was often jealous of these other lovers and told Mabel about one case in which she threatened one persistent woman with an umbrella. Wyatt too was protective of her, though she was less forthright about any dalliances of her own. Their frequent squabbles usually ended in Wyatt putting on his hat and going for a walk.

The book gave Josephine a new lease on life, and the Casons and Ackermans became her new family.

"We had no reason to feel warmth or affection towards her when she came into our lives in her old age but we were all extremely fond of her," Jeanne recalled. "A cute, cute, elfin personality," who always arrived with a fresh cake. She had lost weight in recent years and dressed neatly, usually in black, with her hair combed back and pulled up in a knot. Her face was powdered, her lipstick was skillfully applied, and she had a few treasured pieces of jewelry. "Let me polish up the old furniture," she said before she went out, and then she would check her makeup and tidy her hair. For special events, she wore a white scarf around her neck or one of her beautiful silver fox furs, which she promised to bequeath to the Cason girls someday.

She must have been a striking beauty in her time, Mabel's son Walter Cason observed. Josephine's expressive large brown eyes and her substantial bosom were still her most memorable features; although her figure had lost its sharp curves, her eyes still shone brightly. She was a considerate guest who had to be coaxed to join family photographs and to play in the family poker games. Fastening on a crisp white apron, she prepared excellent meals, specializing in leg of lamb, Swedish meatballs, biscuits, and corn bread. She gave impromptu cooking lessons to the young Cason women and took a great interest in their romantic life. "Our parents had taught us that education came first," Jeanne said, but Aunt Josie was more concerned about whether the smart, ambitious young women would find appropriate husbands, especially Rae Cason, who was studying for medical school. Josephine bought some soft little yellow felt chickens and instructed the girls to tuck them in their bras. "If you wear this, you'll find Mr. Right," she promised.

Josephine encouraged their love of the theater, and when the Cason sisters went to see a performance of *HMS Pinafore*, she waited up to hear everything about the evening and the production. And then she got up and "danced that hornpipe straight through," recalled Jeanne. "You knew without any doubt that she had danced that many times in her youth. I can see her now, her big beautiful dark eyes were sparkling, very expressive, sometimes soft, and you could see the years fall away from her."

The younger Casons often chauffeured Josephine to the Ambassador Hotel for lunch with Sidney Grauman, her old friend from Alaska, or to the MGM lot, where they waited outside impatiently, imagining that she was inside with Gary Cooper, Cecil B. DeMille, or Samuel Goldwyn. Not all the visits were social. Despite her share of Lake's sale of movie rights, and her occasional work as a paid technical consultant, she seemed always to be threatening a lawsuit over the use of Wyatt's name, and was evicted from at least one Twentieth Century-Fox set. The next time Josephine pitched herself as a technical consultant, for a Columbia Studio film titled *The Pioneers*, she was rejected.

The Casons assumed that Josephine had some special connection to the Hollywood moguls because of her religion. In fact, Judaism was a subject of great interest to the well-educated Cason family, who were Seventh-Day Adventists, and they were disappointed that Josephine was cheerfully ignorant. She did like to talk about Jewish food and tell Jewish jokes, and indulged herself in nasty insults against Jews as cheap and low-class. "The worst thing she could say about anybody was that they are just nothing but a kike," recalled Leonard Cason, who remembered her as a "charming funny little old maddening Jew."

The woman who had once relished simple food around a campfire and slept contentedly under the desert skies became a restaurant terror. She could really drive you crazy, Jeanne and her sister Rae recalled. A night out with Josephine usually began with a movie at the downtown Paramount Theater, and ended with dinner at Townsend's restaurant. Josephine would order her favorite: a nice steak, with all the trimmings. But then the torture would begin. "I was an easily embarrassed teenager," Jeanne admitted. "She would order salad, pick through it, wrinkle her cute little nose, call the waitress over, say the lettuce was wilted, or the tomatoes were soggy, point her index finger at it, and say, that salad is not good, I don't want it, take it away! Everybody in the restaurant was dancing a jig to make her happy, and I think she enjoyed the attention."

Josephine's tendency to be suspicious increased as she aged. She feared that people were trying to steal her money, her letters, and the map to her secret mining claims. Harold Ackerman worked from home and had more daily contact with Josephine than was good for either one of them. Where Mabel's husband Ernest was mild-mannered and retreated to his pipe when annoyed, Harold showed his irritation. Resentful of Josephine's presence and her time with Vinnolia, he took so many long walks that his family considered Josephine a great form of exercise.

When she was not staying with the Casons or the Ackermans, Josephine rented a room in downtown Los Angeles or stayed with friends. The Casons were aware that she was in and out of litigation with her family, but Josephine always seemed to have ready cash. Every once in a while, she would change the names on her mining claims, putting them in the name of Mabel's husband or her children. She kept promising Leonard Cason that she would buy him a car when her book was published or when her mines paid off.

Although she was no longer pumping her for information, Josephine continued to visit Allie Earp. It was an unpleasant surprise to discover that Allie was cooperating with a writer. So far, Josephine had managed to keep prying eyes away from Allie, who had the potential to say plenty about her dear friend Mattie Blaylock Earp. Lake had accepted Josephine's excuse about Allie's senility and failed to interview her. Josephine spun a similar falsehood for Mabel and Vinnolia, and mocked Allie as illiterate and ignorant. But now another writer had found Allie. After all this time, Josephine was facing exposure again.

Aunt Allie was well known in her Los Angeles neighborhood, a vigorous walker who marched up and down the streets in her old-fashioned clothes, usually making her last stop in the home of the Waters family, where she would take a rest and put her feet up, a flask of whiskey just visible over the top of her little boot. "We all loved Aunt Allie because she had such wonderful stories and an Irish wit," said Frank Waters. Despite her tiny size and advanced age, Allie was still feisty and ready to tell the story of her beloved Virgil's heroism to anyone who was interested.

Frank Waters was interested in everything about Allie. Wonderful authentic material had fallen into his lap, a big new story from a fresh perspective—and it was all about the famous Earp brothers and the women who loved them. A writer and devotee of Carl Jung and George Gurdjieff, Waters saw Allie as the voice of truth that would validate his preconceived portrait of Wyatt Earp as a publicity-seeking exhibitionist, the ringleader of the criminal Earp Brothers gang, and the representative of the predatory corporate culture that had destroyed the environment and the frontier.

They began to meet regularly. It was easy to get Allie to talk about Virgil, to reminisce about him, and to praise him. But gradually, Frank began to draw her out about Wyatt and the other Earp women.

Waters was loading a shotgun and pointing it at Josephine's heart.

Josephine was unpopular in the Halliwell household; for starters, she was Jewish. In the 1930s, when Los Angeles and the rest of America were experiencing a spasm of anti-Semitism, Allie's niece found no need to hide her contempt. Josephine was "a typical little Jewess," Halliwell declared. Allie blamed Josephine for Mattie's death, and revealed that Josephine and

Wyatt had held back the discovery of the Happy Day mines when they could have shared the news—and the possible wealth—with Virgil and Allie. She did not then know about Josephine's royalties from *Frontier Marshal*, or her list of grievances would have included Josephine shamelessly pumping her for information to feed Lake. "I guess he gypped her too," Hildreth Halliwell concluded, never knowing that Josephine had already received thousands of dollars in royalties.

Under persistent questioning from Josephine, Allie disclosed that she had been meeting with the young writer Frank Waters. There would be none of those stories, Josephine declared. In a minute, they were fighting like the two strong and angry old ladies they were. It would have been difficult to predict the victor of their confrontation. At eighty years old, Allie Earp had more wrinkles but otherwise looked, sounded, and dressed pretty much the way she had in Tombstone. At seventy-eight, Josephine was no longer the doe-eyed teenager in Wyatt's shadow, and she had decades of experience on the public relations battlefield.

Josephine marched right over to the Waters residence and delivered an ultimatum to Frank through his sister Naomi and his mother. "A very nice-looking short aristocratic woman," Naomi Waters observed, despite the surprise attack. Josephine told them that her own book would be published soon, and that she had lawyers on retainer who would sue Waters, as they had sued other authors and movie producers.

After she left, Naomi Waters declared, "The tombstones are rattling all over the place."

"Just one of those women tiffs," Frank dismissed the confrontation. His sister and mother had voiced their support for him, and hence, this "little dust-up" was probably inevitable.

But in the end, Allie wanted to be true to the family. She could shake her fist at Josephine and Wyatt in private, but she would never allow shame to fall on the Earps through her loose talk. The world would not find out about Mattie's suicide or Wyatt's perfidy from Allie. She was an Earp, after all. Hildreth Halliwell came home to find Allie furiously pacing up and down. Frank had read his manuscript to her. It was all a "pack of lies," Allie declared indignantly, and she set aside her feelings about Josephine to form a united front against Waters. Stuart Lake soon joined the fray, since he had no interest in competing for Wyatt Earp readers. He accused Waters of listening to the ravings of a senile old woman who wandered the neighborhood peddling worthless goods from door to door. "Wyatt Earp's widow was a problem in herself," Lake wrote to discourage one publisher, "but she couldn't hold a candle to poor old Aunt Allie."

Luckily for Josephine, Waters failed to find a publisher, perhaps because of the black cloud of litigation that now hung over his manuscript. By the time Waters finally revealed his version of Allie Earp, the two sisters-in-law, who had nothing in common other than their determination to protect the memory of their husbands, would be dead.

JOSEPHINE HAD NEVER wanted to see Tombstone again, but in February 1937 she agreed to a research trip, accompanied by Vinnolia and Harold Ackerman. Mabel stayed behind; Ernest Cason worried that she was exhausting herself with the endless writing and research on top of her responsibilities as an art teacher, wife, and mother.

It was Josephine's first visit to Tombstone since she left in 1882. They drove in the Ackermans' car, with side trips to Dodge City and San Antonio for interviews with some of Wyatt's friends, and a reunion with Albert Behan in Tucson. Josephine found Tombstone sadly diminished, the once crowded streets now lined with empty stores and buildings. They stayed at the Tourist Hotel, which stood on the spot where Virgil had been ambushed. Josephine led Vinnolia and Harold to the house where she had supposedly first lived with Kitty Jones and her husband. They visited John Clum's *Epitaph* office and the old ice cream parlor where Wyatt had tipped his hat to Ann Crabtree and her baby. Bob Hatch's saloon bore a plaque to commemorate Morgan's murder. Josephine inquired of the new owners whether they had known Wyatt, and collected a satisfying raft of anecdotes about the Earp brothers, "fine men, capable and courageous officers who did their duty."

Harold Ackerman drained the last dregs of his patience with Josephine on this trip. Word got around that Mrs. Wyatt Earp was in Tombstone, and as they toured popular landmarks like the Bird Cage Theatre, people gathered around her and asked for her autograph. She managed to alienate even well-wishers. "Everybody is making money off my poor dear husband," she snapped. Her irritability embarrassed Harold, who cringed when she refused to pay a newspaper boy what she considered to be an exorbitant price for a newspaper. Vinnolia found herself caught between placating her husband and calming Josephine. One old-timer volunteered that he remembered "Mrs. Earp" from when she was "on the line," and in case these tourists did not know what he meant, he roared, "Whoring!"

Josephine was all too ready to leave Tombstone again. Unfortunately, forty miles out of town, she discovered that she had left her coin purse somewhere and made them drive back to retrieve it.

Josephine discouraged Mabel and Vinnolia from interviewing some of the people who could have helped with their story. They never interviewed Albert Behan or Allie. They never spoke to Josephine's nieces or nephew, who might have filled in some of the Marcus family background, although they too were keeping secrets: Edna never told her third husband or her own children that her mother's family was Jewish, or that her father had committed suicide.

The manuscript progressed to the point where Josephine began thinking about finding a publisher. She had successfully scuttled Frank Waters's project, but her warning about the imminent publication of her own book was just a good bluff; no one outside Josephine and her writing partners had read any of it.

She consulted Bill Hart first. With his encouragement, she wrote to Harrison Leussler, who was still a literary scout for Houghton Mifflin.

"Really, Mr. Leussler," she said, pitching vigorously, "it is going to be a wonderful book. It ought to be another *Gone with the Wind* because I am telling some history that will cause considerable amazement among the old-timers that are left, and that should make interesting reading for a later generation." Well aware of her previous attempts to censor Lake, Leussler must have been amazed to hear her declare: "I have fully decided to tell the whole story. When you read the manuscript I believe you will be somewhat astonished yourself."

Of course, she did not really mean it.

Josephine fought with Mabel and Vinnolia every step of the way back to the O.K. Corral. Her fabrications about her early life began to confound even these two women who wanted to believe her. Serious researchers, they corresponded with libraries and archives in California and Arizona, determined to verify Josephine's accounts through independent research. For instance, had she actually been part of the Pauline Markham troupe? That much could be substantiated by contemporary references to the McCarthy school, as well as newspaper articles about Pauline Markham and her Arizona tour. They had observed firsthand how Josephine could dance the hornpipe upon request, well into her eighth decade.

But Mabel and Vinnolia could not ignore the mounting number of discrepancies and inconsistencies. Her father was a wealthy industrialist in one conversation, a silk merchant in another, always an upper-class German, never Prussian or Polish, and never a modest baker. She told them that young Josephine Marcus had been captured by Indians and rescued by the dashing young sheriff Johnny Behan. She came to Tombstone to marry Behan and was Albert Behan's governess by day, the houseguest of her friend Kitty Jones by night. When Behan proved faithless, she fell in love with the dashing, heroic Wyatt Earp and flew to his side during the terrorizing moments of gunfire at the O.K. Corral. When she became the lawfully married Mrs. Earp, she was always faithful, never jealous, and only gambled once in a while. Despite her natural tendency to be a "scaredy cat," she was willing to follow her handsome husband into one exciting adventure after another. She adjusted and embellished the story of young Josie to lead naturally to the elegant and socially prominent Mrs. Wyatt Earp, who sounded suspiciously like the elegant and socially prominent Mrs. Emil Lehnhardt.

The other Josephine, "the belle of the honky tonks" or the "two-bit Helen of Troy of Tombstone," as Stuart Lake called her behind her back, was nowhere to be seen. There would be no hint of the existence of Mattie Blaylock Earp. The story of Wyatt's brief marriage to Aurilla Sutherland was fair game, since her poignant death would not compromise Josephine's whitewashed narrative. Josephine had little fear that anyone would discover Wyatt's common-law relationship with the prostitute Sally Earp; Josephine herself may not even have known much about Wyatt's intimate acquaintance with the brothels of Peoria.

As for Wyatt, Josephine's memoir retained just enough of the reluctant gunfighter to live up to the "manhood extreme" image first sketched by Bat Masterson. Her Wyatt Earp was a hero in touch with his feminine side, a domesticated and tenderhearted former sheriff who baked a superior lemon pie and wept over the flowers as he gardened in the moonlight.

Once again, Josephine set out to produce a "nice clean story." This would likely be her last chance to put a twentieth-century halo around Wyatt's violent nineteenth-century past and clean up her own history.

But it was too late: her biographers had glimpsed the real Josephine, the lusty adventurer still lurking within this old lady with the expressive dark eyes.

Mabel and Vinnolia stubbornly clung to their commitment to a truthful memoir. They knew that Josephine's portrait of Wyatt was idealized and incomplete. Harold Ackerman complained that Josephine "doesn't want us to see that [Wyatt] did anything other than teach Sunday school!" They doubted that she had ever been married in any legal ceremony, religious or secular.

They weren't prudes, though: they were open-minded, liberal thinkers, and tried to assure Josephine that they did not judge her. They slowly drew from her some of the truth: She had run away from home to become an actress, and Johnny Behan wanted more from her than babysitting for Albert. Wyatt had been a gunfighter, a gambler, a saloonkeeper, and a womanizer, before and after he met Josephine. They sympathized with her fear of exposure, but they could not convince her that some unspoken statute of limitations had run out, and that she would not be vilified for any of these transgressions. Even when they confronted her with some of the real history they had uncovered, Josephine would say, "I don't want to use that story; it makes me look like a bad person." And, of course, Mabel and Vinnolia still knew nothing about Mattie.

"We finally came to realize that [Josephine] was not giving us a vital part of her story, although we had a finished manuscript. She was not telling what happened at Tombstone or how she met Wyatt. Because of her actions we realized that something had happened at Tombstone that she simply would not reveal," said Mabel's daughter Jeanne.

The manuscript was "finished," but it bore all the hidden scars of something written with an uncertain voice and a hole in the middle of its heart.

Josephine and the sisters selected their favorite chapter and sent it, complete with illustrations by Mabel, to Harrison Leussler at Houghton. His response was guarded. "I like the style of the writing," he said, and complimented Mabel's original watercolor drawings. Overall, he noted that the manuscript showed considerable promise. However, he would have to wait and read the whole thing. "I have no doubt that you are writing an interesting story, but it is just simply a case of finishing it. You are perfectly right in telling the story of your life and I do feel that there will be enough interest in your own experiences to make this an interesting volume," he ended somewhat coldly.

Josephine called his response "a mixed up affair" but she thought they had received at least a blinking green light from Leussler and Houghton Mifflin.

Unbeknownst to Josephine or her cowriters, Leussler immediately sent their draft chapter to Stuart Lake. His motivation was probably to protect future sales of *Frontier Marshal* and keep its author happy, but his action was premature at best, and Mabel later considered it underhanded. Of course, Lake wanted the Cason-Ackerman manuscript buried at the bottom of every publisher's slush pile.

Ultimately, though, it was neither Lake nor Leussler who sabotaged Josephine's planned memoir, but Josephine herself. The battle over what was in and what was out of the book had escalated after her return to Tombstone. The same arguments between biographers and subject went back and forth: Wyatt must not look bad; Josephine should appear to be a respectable woman; tell the truth.

After one particularly stormy session, Josephine declared that she was finished. She wanted the manuscript destroyed.

The fire was prepared. Josephine Earp watched Mabel Cason feed papers into the flames. One by one, hundreds of typewritten pages turned into ashes. As the pages were consumed, Josephine evoked a curse against anyone who would disturb the secrets of the past.

Mabel and Vinnolia had spent years preparing this manuscript. They had fed and housed "Aunt Josie" and carefully researched and cheerfully ghostwritten and coaxed and edited and followed her whims—and now they were destroying four years of work at her command.

They agreed to abandon the project, but they had no fear of her hex, which they took to be some kind of Jewish superstition. They had invested so much time in the project that they believed they had a perfect right to trick her. Some of what they had written was based on their own research, not interviews with her. Each of them kept copies of their sections, as well as their notes and rough drafts. Safely stored, unbeknownst to Josephine, the secret manuscript would the "last chance for history," they felt, and should be preserved for some future purpose.

There would be no hard feelings, however, at least not from Mabel and

Vinnolia or their children. They remained sincere in their affection for Josephine. Their husbands were less forgiving. Ernest Cason regretted the endless time that Mabel had spent on the dead manuscript and declared that henceforth, "Wyatt Earp was a dirty word." Harold Ackerman had an even longer list of grievances. When Josephine accused his son of stealing from her (the money was misplaced), Harold found her an apartment and moved her belongings. Vinnolia protested to no avail. Harold agreed to store some of Josephine's things in a trunk in his basement, but the woman herself had to go.

Josephine would spend her last years with friends or in cheap rentals. She kept in touch with Flood and Mabel, and had long visits with Sarah Lewellen, her old friend from Tombstone. She was friendly with her landlady, although she was also usually behind in her rent. She had some new friends, Wilfrid and Bessie Nevitt and the Sinclair family, who were kind to the increasingly forgetful and needy old woman. None of her friends, old and new, seemed to know each other; she seemed to prefer that nobody had the full picture of her life.

Edna relented and made a small settlement upon her aunt, paying her retroactively for some of the years that the ownership of the oil wells had been in dispute. Nevertheless, Josephine continued to borrow money. John Flood remained a frequent target. Millionaire Ed Doheny put her in his "Dead Beat" account book, with standing orders to his staff not to admit her again. Leussler and Lake exchanged sarcastic notes on how much they would allow her to borrow. Lake recommended her for a job as a film consultant on a film, *Belle of the Yukon*, perhaps out of pity, perhaps to ease her growing demands for advances on their shared royalties.

She was now almost the age that Wyatt had been when he died. She had lived without him for more than ten years. Like the two women she had met so long ago, the maid in Salt Lake City and Baby Doe Tabor in Denver, she had ended up childless and penniless, protective of her husband to the bitter end. She was a frail old lady, weighed down by a large black knit purse that she always carried. At the least sign of interest, she would snap open the purse's large shell-shaped catch and bring out a stack of newspaper clippings, crumpled and worn with folding and refolding. Beneath them could be glimpsed one hundred bills wrapped in toilet paper, and yellowed calling cards engraved "Mrs. W. Earp."

The clippings were not about Wyatt Earp, however, but about Hattie Lehnhardt. "My poor dead sister!" she exclaimed, showing a visitor the articles about her sister as chief executive of Lehnhardt's of Oakland, a business and society lady, presiding over formal lunches and opening night at the San Francisco Opera.

When Harold and Ernest permitted a visit, she returned to Mabel and Vinnolia, who were still fond of her, as were their daughters. Rae had a new beau, a devotee of *Frontier Marshal*. Josephine warmed to the intelligent and sincere young man immediately, but spent the afternoon talking about the Marcus family, her poor dead sister, and the accomplishments of her nieces and nephews, with nary a word about Wyatt to the disappointed suitor.

Her private reticence did not translate to any relaxed vigilance when it came to Wyatt's public reputation. Reading an announcement of a forthcoming article about Wyatt Earp in *American Weekly*, she asked Flood to write once more under her signature to William Randolph Hearst. "Will you please be so kind as to halt the story, until it has been well advised and that I be granted consideration," she requested. "Such would be in keeping with the regard of your father for Wyatt Earp and the fondness of Wyatt Earp for your long remembered father. With all this, I am carried back, in memory, to the perilous days of Tombstone where the firm friendship of the one for other developed."

Perhaps it was this effort that spurred her to make one last run at publishing her story. With Flood's remarkable forbearance, she commissioned a series of letters to a short but powerful list of potential collaborators, including film connections such as Sol Wurtzel of Twentieth Century-Fox and Edward F. Cline of Universal Pictures. To Walter Coburn, who had previously been on the short list as her collaborator, she wrote, "For several years past, I have been gathering together material for a story, somewhat in the nature of a biography." Coburn declined again, citing other commitments. Josephine went on to make inquiries of William S. Hart as well as her old friends at Houghton Mifflin. Her last hope—and one that cost her much in the way of dignity—was Stuart Lake: "Perhaps you have good news for me; that you are prepared to commence work on my story. I trust that such is the situation, and that I shall not be disappointed longer, nor my friends either."

Had the timing been different, and had her health and memory not been an issue, Lake might have finally written another book about Wyatt Earp. His editors, despairing that Lake would ever finish his long-promised book on Wells Fargo, encouraged him. Josephine had taken to calling him at odd hours, and he found her "lonely and forlorn." Some unknown person had conned her out of most of her nest egg, and by the time Lake discovered the theft and hired an attorney on her behalf, the damage was done. "I know who turned the various tricks," Lake said, but never revealed the culprit. He had not lost his interest in the story that he had been denied so many years before. With his old bravado, he wrote to Houghton Mifflin that he finally had the real story of Tombstone. "Within the last year or so Mrs. Earp broke down and came clean with the whole works." It would be the basis of a new film, he promised.

Lake would never write that story.


IN 1937 JOSEPHINE named John Flood as her executor, as well as the beneficiary of the polar-bear-skin rug and the Happy Day mines. In this version of her will, she left the copyright and future royalties from her book *My Life with Wyatt Earp* to the family of Mrs. Addie Schofield Sinclair. Her nieces and nephew and their children were to receive equal shares in her mines. As to the future royalties from *Frontier Marshal*, she left the beneficiary's name blank. Forgetful or angry or uninterested in the outcome, she destroyed that will, leaving only Flood's carbon copy in his files to reveal her intentions.

In 1944, she owed money all over town, and her rent was overdue. She was living at the home of Mrs. Lon Maddox, at 1812 West Forty-Eighth Street, her third or fourth address in Los Angeles since Wyatt's death. Her total property included one trunk, one radio, and five boxes of miscellaneous items, estimated at a total of $175. Another trunk remained at the Ackermans' house, and a third trunk had been removed from the Oakland mansion before Hattie's death and stored in Edna's home. She had been selling or giving away whatever would buy her goodwill or a cushion of cash. Her niece Alice had two of Wyatt's canes and the bearskin rug that had been a gift from William S. Hart. The Casons received some monogrammed linens, a Bible inscribed to Wyatt, and a cameo of Josephine. Those beautiful silver fox furs coveted by Jeanne and Rae ended up around the neck of her lawyer's secretary. Perhaps she had simply forgotten her promise, or perhaps she owed a debt that she repaid with the furs.

She composed a new will, just one short paragraph, shaky in syntax and skimpy on details:

> *I Josephine Earp in sound mine am sick, and if anything should happen to me I want Bessie Nevitt, to take full charge of all my belongings. She will know my bags copy my full life story will protect picture made of my late husband Wyatt Earp fight for me just if I would do she and her husband Wilfred Nevitt will have full charge and will do just like I want to do. These are my wishes in sound mind.*

"I know you're in there, Mr. Flood!" In the last year of her life, Josephine had taken to sliding angry notes under John Flood's door when he was not home, or when he chose not to open the door. Her aggressiveness frightened him, and he began to keep track of her visits with handwritten notations on the back of calendar pages, with quotes from her: "I'll get back at you—good and hard." Finding his door locked, she rang the bell and banged on the door. Once she stuck her arm through the screen door to reach the doorknob. He recorded her threats—"I would not open it either if I were you"—in faint pencil on the back of another calendar page.

Flood's calendar pages kept growing. He added a few random dates and Josephine's address, the names of her landlady, her attorney, and some of her debtors. Her bizarre visits stopped sometime in the fall of 1944.

Josephine died on December 19, 1944.

A terrible silence followed her death. No immediate family stepped forward to make the arrangements for her funeral; it was William S. Hart and Sidney Grauman who paid for the brief service at the Armstrong Family Mortuary, but neither of them attended. Rabbi Maxwell Dubin officiated, most likely pressed into duty by Sidney Grauman. Her nieces were absent. Mabel and Vinnolia were unaware of her passing for several months. Cremation followed immediately at Westwood Memorial Park. Flood stayed away, though he received a copy of the program from one of Josephine's friends, who chose the 1936 A. L. Frank poem "The Rose Still Grows beyond the Wall." Her ashes were later sent to her niece Alice Cohn, who brought them to the Hills of Eternity cemetery, where Josephine would come to rest next to her mother, father, brother, and Wyatt.

It was six months after D-day, the Battle of the Bulge was still raging, and the British were pushing forward in Burma. Despite a world at war, there was room in the news for a small obituary to run in the *Los Angeles Times* on the death of the "widow of the picturesque western frontier gunfighter, United States Marshal Wyatt Earp." Cause of death was cited as a heart attack; her death certificate also listed dementia as a contributing factor. While her passing was noted in some of the places where she and Wyatt had lived, there was none of the outpouring of affection or reflection that accompanied Wyatt's death.

Her will was never properly executed. Bessie Nevitt sued unsuccessfully for probate soon after Josephine's death, and then several creditors began to press claims against the estate. Her landlady retained her personal effects because she had not paid the rent. Flood failed to collect his loan of $157.

Lake called on Josephine's family and pressed them to pay her debts to him. He was also searching for a manuscript that she had left behind, presumably in association with their proposed collaboration. Niece Alice Cohn turned the question around to Lake. Having read that *My Darling Clementine* was in production, she pressed Lake for information about whether Josephine's manuscript might have been stolen by Bessie Nevitt or Mrs. Maddox. Surely the consent of the family was required for any motion picture sales, she argued, and Josephine's three nieces were her rightful heirs. As they were soon to find out, however, Lake had no intention of sharing future movie rights payments or royalties with them, or with anybody else.

Writing in 1952 to an Earp relative who had inquired about Wyatt's place of burial, John Flood filled in the blanks: the circumstances of Wyatt's and Josephine's death, and the names and addresses of Josephine's living relatives.

"I am a bachelor," he noted, "[residing] in a small cottage in the rear of 2020 Fourth Avenue. . . . When you are in Los Angeles, be sure to call. It, probably, will require several months to tell the story although, I promise it will be thrilling enough so that you will not go to sleep while listening."

# 6 | PLANET EARP

*W*YATT EARP and the Gunfight at the O.K. Corral never went out of style. Almost anything based on western themes and events continued to grab the public's attention, and making movies and publishing books only became easier after Josephine's death. The facts about the real person faded away. There were no visitors to a certain Jewish cemetery outside of San Francisco, because Wyatt's burial place was unknown.

Nobody was looking for Josephine.

Not even a world war stopped the interest in Wyatt Earp. Although there would be dozens of biographies, long and short, scholarly and speculative, *Frontier Marshal* held its position as the most popular. During World War II, some 100,000 American soldiers received free paperbacks, courtesy of the Council of Books in Wartime.

Lake no longer had to share the money. Ever wily, he convinced Josephine's family that they should not expect any royalties, since the era of Earp was over, and besides, Josephine died in debt to him. Lake was disingenuous at best: there were several new Earp-themed projects in the

Hollywood pipeline, and a newfangled thing called television was about to transform the media world. Wyatt Earp was headed to the next stage of immortality, and Lake preferred to ride alone.

Tombstone's hero burst into America's living rooms on September 6, 1955. *The Life and Legend of Wyatt Earp* aired on ABC, with handsome Hugh O'Brian playing the role of Wyatt Earp and a theme song so memorable that it would linger in the audience's head long after the show went off the air, six years and 249 episodes later. Lake wrote a new edition of *Frontier Marshal* for teenage readers and served as a consultant for the creation of the television show and its first thirty-five episodes. The story lines were lame, and most episodes featured gunshots so loud and frequent that O'Brian suffered permanent hearing loss. But it was one of the most popular shows on television, shortly followed by *Gunsmoke* and *Bat Masterson*. Soon every major broadcast network had a production tied to Wyatt Earp.

Josephine's relatives and friends tuned in, too. The show reminded her great-grand-nephew Felton Macartney that he had seen some old steamer trunks in his mother's basement labeled "Property of Wyatt Earp." To his disappointment, he found there only stinky old clothing, an address book, and a watch fob.

OF THE PEOPLE who had been involved with Josephine in the 1930s and 1940s, a few were still alive. Harold Ackerman was now a wealthy widower who watched a lot of television. He wrote to Mabel how he missed Vinnolia, who died in 1954, and how he wished they could all watch *The Life and Legend of Wyatt Earp* together.

Mabel hoped some day to finish Josephine's story. She thought she had found a new coauthor when her son Walter introduced her and Harold Ackerman to his former classmate, John Gilchriese, a devoted historian and collector of all things western. Gilchriese's sincere interest in frontier life drew the dwindling number of pioneers to him. With a photographic memory and a deep grasp of detail, he remembered the names of their sons and daughters and grandsons and granddaughters, and in gratitude, they spun their yarns about life in the Old West, and pressed souvenirs and clippings on him. As he traveled all over Arizona, his collection grew from first editions of Jack London and Mark Twain to precious manuscripts and maps, and when he ran out of space in his small house, he rented garages and storage spaces. Eventually, Gilchriese would claim a deeper connection to the people and papers associated with Tombstone than anyone else—and he had the goods to back up his claims.

Now that Wyatt was the hero of the American living room, Harold was angry all over again about how Josephine and her "shrewish" ways had cheated Mabel and Vinnolia out of their rightful place as narrators of this untold part of Wyatt's story. "You can't tell me Wyatt was a killer," Harold declared. "He lived with Josie for nearly fifty years!"

NO ONE HAD yet told the story of love at the O.K. Corral. Mabel saw it as a unique woman's story and wanted to claim a special place for Josephine. She felt the pressure of history and a specific fear that Lake would scoop her yet. "I'd like to be the first to give it to the public," she said. "I think my sister and I earned that right. And [Lake] got it in a manner, by fraud it seems to me."

Harold begrudgingly admitted the value of Josephine's perspective, since "the estimate of a man told by his wife of half a century should have some element of truth." He encouraged Mabel to sign a formal partnership agreement with Gilchriese.

Although she was a far more experienced author than Gilchriese, Mabel took the young man at his word that he could fill in the big holes left by Josephine's misdirection, especially about Tombstone. She set to work again with her previous manuscript and her voluminous research notes. She knew Josephine better than anyone alive. Who else could speak authoritatively to her history and her distinctive character traits, right down to her "Brooklynese" accent? "I want to give quite an accurate picture of her without making it too derogatory—she had her good points which can be emphasized," Mabel said, sketching out her portrait of Josephine in advance for Gilchriese. "She had quite a violent temper and was jealous. I'd like to bring that out too. She also said that in their prime Wyatt was jealous of her." But she knew that she would be stymied once again by the mystery of Tombstone, and looked to Gilchriese to "prime" her on that critical year. "My material is untrustworthy—and perhaps dangerous," she admitted. "My bombardment of you with letters begins with this," she wrote on October 6, 1955, and posed a list of the questions that most tantalized her: Did Josephine dance at the Bird Cage Theatre? Was "Little Gay" her stage name? She hoped that Gilchriese would help her locate Josephine's family.

They would have been a great team: Mabel's psychologically rich portrait of Josephine set like a jewel within Gilchriese's comprehensive knowledge of the era, with her elegant, clear prose bringing Josephine back to life. Mabel worked "day and night" on her draft, fretting about whether the reader would be impatient if she took two whole chapters to get Wyatt and Josephine together in Tombstone.

Gilchriese, however, shared nothing with Mabel. She knew nothing about his other sources, and would have been amazed to discover that John Flood, Wyatt's friend and first biographer, was still alive and in possession of a treasure trove of original documents, as well as a deep source of information—and prejudices—about Josephine. Together, Flood and Gilchriese had walked the streets of Tombstone. For the orphaned bachelor Flood, the Earps had been the most meaningful relationship of his life. Still angry that Lake had exaggerated his own closeness to Wyatt Earp, Flood saw Gilchriese as the path to restoring his rightful place. Of course, he had no idea that Gilchriese also knew Stuart Lake and was now acting as double agent with all of them: Cason, Lake, and Flood.

Recent years had elevated Wyatt in Flood's estimation and lowered

Josephine. He apparently never wrote down his stories about her but shared many of them with Gilchriese. What Flood had seen or heard firsthand, and what Gilchriese invented, would remain unknown. However, relying on stories that were purportedly told by Flood, Gilchriese certainly devalued Josephine's stock still more among Earp loyalists and enemies alike. He was the source of the story that Josephine had exploited the Cason and Ackerman families for food and shelter during the difficult last years of her life, an accusation rejected by Mabel's family. Josephine "came with the intention of writing the book, not trying to mooch off them," Jeanne Cason Laing said. "My mother was sick to death of Mr. Gilchriese and his little tricks."

His was a vindictive, imperious, and bitter Josephine: everyone disliked her, but some disliked her more than others and nobody cared when she died. Gilchriese attributed to Wyatt's niece Alice Earp Wells the theory that the Earps' fifty-year relationship was held together not by affection and shared history but by some dark secret that Josephine knew about Wyatt. "Get your ass in the kitchen," Wyatt would snarl, a pattern that would kick off a screaming match, with Josephine hissing at Wyatt about shortcomings ranging from his sexual prowess to his breadwinning capacity, always throwing Johnny Behan in his face as the rival she should have stuck with. Josephine kept Wyatt isolated from his friends and family, and refused to allow him to see his mother or father right before their deaths. Josephine's compulsive gambling wrecked their finances and left heavy debts all over town, driving a wedge between Wyatt and old friends like Ed Doheny. Josephine was obsessed with money and with her appearance: she refused to attend Wyatt's funeral because she did not have proper clothes, complaining anew that Johnny Behan would have left her enough money for the damn funeral and a dress. When her husband's container of ashes was brought to her, she smashed it against the wall and then in a fit of rage, she swept the ashes off the carpet and out the door. Whatever was eventually buried in Colma was NOT all of Wyatt Earp.

But the most unkind accusation of all was that Josephine was unloved.

Gilchriese asserted that Wyatt was infatuated with a married woman in Tombstone, and that he continued their secret affair in Los Angeles. This unnamed woman was beautiful—much more beautiful than Josephine— and Wyatt was much fonder of her than he was of Josephine. The evidence of this was supposed to be a formal portrait of the two of them taken in the famous photography studio of C. S. Fly, inscribed on the back by the mystery woman to Wyatt, and then given by him to Flood, who gave it to Gilchriese.

Gilchriese shared these stories with other people, but not with Mabel. However, she had already figured out that this new writing partnership was a mistake. She was learning nothing. In her opinion, Gilchriese had turned out to be "a bag of wind." She warned Harold that they had been taken by Gilchriese, who was a "good salesman" but a poor writer. "I don't see that he has a bit of material that we do not already have or can get easily," she wrote to Harold. As far as she was concerned, Gilchriese had broken their agreement.

Without showing her draft to Gilchriese, she sent it to Bobbs Merrill Publishing Company, and it was rejected.

Mabel sent Gilchriese a curt and final dismissal of their relationship: "In view of [the publisher's rejection] and in view of the fact that you do not seem to have material on the Earp story that we do not already have, we will forget about the book on the life of Josephine Earp."

At first, Mabel hoped to go forward on her own. She was not deterred by the publisher's rejection and she was not afraid of Gilchriese, not even when he threatened a lawsuit and warned that Lake would also prosecute her. Writing at the end of 1955 to her sister, she signed her letter "Mickey de Tough," and dispassionately analyzed the pros and cons of proceeding alone. "I hate to lay down a job I've started," she said. She felt that her legal position was strong, both with Gilchriese and with Lake.

But Mabel had little appetite for the protracted strife of a lawsuit. The stress on Harold Ackerman was already showing; he was shocked by the betrayal of his young companion Gilchriese. "Poor Harold—he is so fearful—afraid of life—afraid of death—afraid of men and afraid of himself—it is a pitiable state of mind," Mabel said to her sister. Was it worth it? "As a Christian I wonder if it has enough of constructive value for me to spend my time and questionable talents on—on the other hand I can sometimes believe that there is something of constructive value in it."

Mabel consulted with her husband, who advised her to set aside the manuscript for the second and last time, and suggested that they should have prayed more before signing the contract. "Don't mention the words Wyatt and Earp when you visit—them's fightin' words with him," she warned her sister.

Sadly, 1955 was looking a lot like 1938, as if Josephine had renewed her curse.

Nineteen fifty-five was also the year when the truth about Mattie Blaylock Earp finally emerged.

The thunderbolt came from Kansas. An article about the opening of a new Dodge City museum reminded a Mr. O. H. Marquis from Iowa that he had seen pictures of young Wyatt Earp in a trunk that he inherited from his "Aunt Ceely." Marquis contacted the museum, which authenticated the photos and a Bible inscribed to Wyatt by the grateful citizens of Dodge City.

When the news reached Mabel, she understood that Josephine's worst nightmare was about to burst into daylight. The whole world would soon know that "Aunt Ceely" was Celia "Mattie" Earp, and that Wyatt and his lover Josephine were the cause of Mattie's degradation and death.

"That is what Josie was covering from us," Mabel understood. "She seemed to be truly conscience-stricken about it." It would have been little comfort to Josephine to know that Mattie herself blamed Wyatt most of all.

If Mabel had any doubts about the harsh truth of Wyatt's infidelities and Josephine's role as the other woman, they were removed by corroborating documents from Tombstone, including the mortgage that Wyatt signed with Mattie Blaylock as well as *Epitaph* clippings that bid farewell to "Mrs. Wyatt Earp" after Morgan's murder. "Wyatt Earp seems to have been a very brave man and a stickler for the law," Mabel noted with some asperity, "but his personal life seems to have left much to be desired." This was what Josephine meant when she sighed that Wyatt "didn't have the best of principles where women were concerned." Then again, for generous Mabel, this new realism in the cult of Earp was inevitable when "humans were misrepresented as heroes, instead of ordinary men."

The discovery about Mattie forced a reinterpretation of everything about Josephine and Wyatt. As Mabel had predicted, Lake's saintly image of Wyatt Earp was about to be stained.

More debunking came about with Frank Waters's long-delayed publication of *The Earp Brothers of Tombstone* in 1960, which arrived in time to match the national mood of cynicism and mistrust over the coming Vietnam War. In his view, it was not enough for false heroes to be made human: the old icons had to be utterly smashed. Stuart Lake had written a "fictitious legend of preposterous proportions," America's most hateful "morality play," and the source of our "tragic national psychosis." The real Wyatt Earp had spawned America's materialistic ideology, the pop culture expressions of pulp fiction, radio serials, toy pistols, and tin badges. Waters inveighed against the destruction of the environment, the contamination of the water supply, and ultimately, the degraded quality of human life and the American obsession with "the omnipotent dollar, economic and police corruption on all levels, and the streak of violence imbedded in our nature as a people."

As for Josephine Earp, her sins were more modestly carnal. Waters found her guilty of nothing more cosmic than breaking Mattie Blaylock's heart. He used Allie Earp to reimagine the romance of Wyatt and Josephine as public debauchery that ground vulnerable Mattie into the dust of depression, from which even her loving sisters-in-law and a sympathetic Big Nose Kate could not rescue her. He told of bitter fights between Mattie and Wyatt, and Mattie's humiliation at the spectacle of Sadie Marcus, the slut of Tombstone, flouncing along the streets of Tombstone. All of this was dramatized in Allie's irresistibly folksy dialect:

> *"We all knew about it and Mattie did too," said Allie. "That's why we never said anything to her. We didn't have to. We could see her with her eyes all red from cryin', thinkin' of Wyatt's carryin' on. I didn't have to peek out at night to see if the light was still burnin' in her window for Wyatt. I knew it would still be burnin' at daylight when I got up. . . . Everything Wyatt did stuck the knife deeper into Mattie's heart. Polishin' his boots so he could prance into a fancy restaurant with Sadie. Cleanin' his guns to show off to Sadie. You never saw his hair combed so proper or his long, slim hands so beautiful clean and soft."*

Allie, however, had said nothing of the kind.

It would be decades before Waters's portrayal of Allie would be exposed by resourceful researchers and historians as "hogwash," as author and historian Casey Tefertiller later put it, the charred remains of a torched Earp legend that was every bit as distorted as Lake's hagiography.

In a letter to the Arizona Historical Society, Waters later admitted that he had combined Allie's words with a "cold, objective analysis" and "expose" of the whole subject. However, he influenced a generation of readers who thought they had heard the true voice of Allie Earp, and adjusted their views to accommodate the figure of Josephine as the heartless hussy of Tombstone with Wyatt as her villainous consort. These were the new legends of Tombstone.

AMONG THOSE ENRAGED by the new anti-Earp dogma were Wyatt's grandniece Estelle Josephine Miller and her husband Bill Miller. Despite having been named after Josephine, Estelle had to struggle to remember anything particularly good to say about her aunt. Wyatt, though, was entirely different: she remembered him as considerate and friendly, a gentle giant who bounced her on his knee. "My Uncle Wyatt wasn't like them writers say," she insisted. He was neither Boy Scout nor Saint Wyatt the Just, but a "rough, tough, profane, rooting-tooting frontiersman." They consulted with an interesting writer of their acquaintance: What could be done to rescue Wyatt Earp from the mud that threatened to sink him forever?

Glenn Boyer was the writer, a dashing Air Force pilot with a wide iconoclastic streak and an obsession with all things western. His interest in Wyatt Earp began with a 1937 article in the *Chicago Tribune* titled "Tombstone, the Town Too Tough to Die," which led him to *Frontier Marshal* and a brief correspondence with Stuart Lake. Boyer had a good nose for research. When military duty brought him to Earp-rich locations such as San Bernardino and Colton, he haunted local archives and chatted up the old-timers, and eventually found his way to the Millers.

With the Millers' encouragement, Boyer began to think about writing a more balanced, nuanced portrait of Wyatt Earp. This would be his second book; the first was a slim pamphlet called *The Illustrated Life of Doc Holliday*, the cover of which promised "Sensational Photo Discoveries." He delivered on that promise, with previously unpublished documents and photographs that were indeed sensational, especially a newly discovered letter from Doc to a friend identified only as "Peanut," in which Doc Holliday admitted that he and Wyatt Earp had killed two men in Colorado and buried them under rocks.

Boyer was recognized as a major new voice in western history. In his next book, *Suppressed Murder of Wyatt Earp*, Boyer put forth a stylish and entertaining argument that the historical Wyatt Earp had been supplanted by equally misleading caricatures: a plaster saint or grotesque villain. With Boyer's apparent dedication to telling the truth about Earp, the wildly swinging pendulum seemed to have reached a new equilibrium and identified a new champion for Josephine and Wyatt.

Boyer next contacted the Casons to propose a book based on their material. Ernest Cason responded with the sad news of Mabel's death in 1965. Despite his oft-expressed distaste for anything to do with the Earps, he verified what Mabel had learned about Mattie's life. Ernest's interest was in fidelity to the facts and he dismissed Stuart Lake's book with the comment that it should be "considered fiction rather than history."

Boyer's approach was well timed. After all, Jeanne Cason Laing told her father, Boyer was a colonel in the Air Force, "not just some bum off the wayside." The Cason family admired Boyer's previous books and agreed that Boyer would be a worthy choice to finish Mabel's work.

With Ernest's approval, his daughter Jeanne Cason Laing scoured her mother's files and sent them to Boyer. It was "a stack of material almost a foot high" that belonged to the three different eras of Mabel's immersion as Josephine's biographer: drafts and notes from Mabel's and Vinnolia's first manuscript in the late 1930s, which included something that Jeanne called "the Clum manuscript"; materials and letters that related to the 1955 attempt, including correspondence with Gilchriese and Houghton Mifflin; and clippings and correspondence with researchers during the mid-1960s. "You and my mother and my aunt would have liked each other," said Leonard Cason, echoing the warmth that the rest of the family felt toward Glenn Boyer.

Over the next four decades, Jeanne Cason Laing and her family would have reason to reconsider whether they had chosen the right candidate to complete Mabel's mission of truth.

JEANNE CASON LAING became the main liaison to Boyer. The whole family was a bit jaundiced about the subject, she warned him, and she bristled immediately at his clumsy suggestion that her mother, who was known as Mabel Earp Cason, had traded on the name Earp. It was simply her maiden name, a choice that Jeanne too had made. "There are many more worthwhile Earps than Wyatt—and I count my mother among them," she scolded him. If anything, "the whoop-de-do about Wyatt has cheapened the name."

Boyer recovered from this misstep and for the next nine years, Jeanne gave him exclusive access to the original manuscripts and to the family's still vivid memories of Aunt Josie. He also tracked down Josephine's living relatives, sometimes accompanied on his visits by an enthusiastic Jeanne Laing. The women laughed at his banter about their sex lives and physical appearances, while he talked about shotguns and shootouts with the men, all the while advancing his research agenda with reasonable grace. "If you happen to run into a casual picture of Wyatt as he is headed out for a stroll, please don't give it to the Salvation Army," he instructed his interview subjects, who promised to be on the lookout for documents, photos, and memorabilia.

With the help of Mabel's research materials, interviews with the Miller, Cason, and Marcus families, and his own independent research, Boyer published *I Married Wyatt Earp* in 1976 with the University of Arizona Press; the title had been originally suggested by Mabel and Vinnolia. For sources, Boyer leaned heavily on the Cason manuscript and interviews with the Millers and the Casons, supplemented by "the Clum manuscript," documented by an entertaining set of footnotes. On the book's cover was a nearly nude figure, her voluptuous figure thinly veiled, her brown eyes framed by a fall of dark brown hair. Boyer identified the woman as Josephine, who posed for the provocative photograph at Johnny Behan's request.

A great title, a lively story, and a racy cover goosed the sales and Boyer's reputation. His Josephine was neither Waters's whore nor Gilchriese's materialistic dominatrix. She was an energetic, affectionate, and good-humored woman, her sentimentality balanced by a bottomless appetite for adventure and a love of nature. The Casons were pleased. The sisters requested autographed copies and complimented the author on "the most truthful account" of the Earp saga. "We're so glad to see all of Mother's and Aunt Vinnolia's work brought to fruition," they wrote. Josephine's family was also enthusiastic. Any historian's heart would have beaten faster to hear about a sheath of typewritten pages found by Josephine's grandniece Alice Greenberg, together with what appeared to be drawings of the Gunfight at the O.K. Corral. This turned out to be the only two copies of John Flood's typed manuscript, with his original sketch of the gunfight, drawn from Wyatt's memory. Boyer purchased and published the manuscript as a commemorative edition in 1981, the centennial of the gunfight.

Now he had it all.

Boyer kept up a steady stream of publications and appearances after *I Married Wyatt Earp*, and also stayed in touch with the Cason and Marcus family. It was, however, one of the rare dips in the popularity of Earp-themed films, and Hollywood did not come a-calling. Instead, Boyer received a film credit on a low-budget television movie starring Marie Osmond as Josephine Earp.

Questions first arose about the authenticity of Boyer's work when his book on Doc Holliday came under attack. His great discovery, the "Peanut" letter in which Doc Holliday confessed to another murder, was no scoop at all, just a rollicking good bit of creative writing. Eight years after its publication, after at least one unsuspecting historian relied on its findings, Boyer declared that the early work was a deliberate spoof, a "wild story" intended to entertain readers and entrap sloppy writers. He released it again with a new advertisement that admitted its fictional roots. However, sharp eyes were watching.

With Boyer's next book, *Tombstone Vendetta*, he flaunted his indifference about whether his critics found his research to be good enough. Almost immediately, the book was accused of being another fictional flight. The sins of *Doc Holliday* and *Tombstone Vendetta* were then visited on *I Married Wyatt Earp*: if Boyer fooled us twice, perhaps he had also bowdlerized the memoirs of Josephine Earp.

Boyer lobbied the University of Arizona Press with a blizzard of letters that proclaimed the historical authenticity of *I Married Wyatt Earp*. But the Press was looking worriedly over its shoulder at Boyer's noisy and persistent critics. The lawyers threatened to withdraw the book unless they were allowed to label it "historical fiction."

The legitimacy of *I Married Wyatt Earp* took another blow when that provocative cover photograph of Josephine was alleged to have been mass-produced around 1914, long after Tombstone.

"If it isn't Josie, it ought to be," Boyer grumbled.

Under pressure from indignant historians and journalists, the Press gave up all rights and returned its unsold copies to Boyer, who promptly announced that he intended to write his own account of the scandal, to be called *I Divorced Wyatt Earp*.

Boyer lost the ability to see his seminal work influence the next generation of Earp writers. If he believed that nobody could produce major new works without access to his knowledge, or at the very least, without some citation as a source, he was wrong. In 1997, Casey Tefertiller published *Wyatt Earp: The Life behind the Legend*, a comprehensive and scrupulously researched biography that almost completely omitted Boyer, turning back to original source material and also cultivating the author's own relationship with the Cason, Marcus, and Welsh families. Boyer was also invisible in Allen Barra's *Inventing Wyatt Earp: His Life and Many Legends*, and from Barra's bully pulpit as a respected and prolific freelance journalist for publications that included the *Wall Street Journal* and the *New York Times*, he kept up the heat on Boyer's credibility.

In 1990 the community of Earp writers and historians was galvanized by the announcement of two major Hollywood productions, *Tombstone*, starring Kurt Russell and Val Kilmer, and *Wyatt Earp*, starring Kevin Costner. Fame and fortune were on the line, as Earp experts vied for lucrative contracts and the all-important bragging rights that went with being official consultants on a major Hollywood production.

Boyer was completely shut out of the biggest and most exciting game in town: the dueling arrival of two major motion pictures about a subject that he believed he knew better than anyone else in the world.

JOSEPHINE WAS USED as a punching bag by all sides of the *I Married Wyatt Earp* controversy. Over the years, memories grew dimmer, tempers shorter, and the authenticity of original Earp documents more hotly contested: experts and amateurs jousted about the Flood manuscript, Allie Earp's memoirs, the transcript from the 1881 Tombstone hearing, the letters of Louisa Earp, the manuscript of Forrestine Hooker, the maybe-but-not-for-sure diary of Wyatt's niece Adelia Edwards, the papers of Big Nose Kate, and a Grand Canyon–size collection of miscellaneous guns, recordings, beds, scraps of cloth, and photographs of questionable provenance. But none caused more grief than the so-called Clum manuscript.

The idea that John Clum had written an early, unfinished version of Josephine's story originated with Mabel's daughter Jeanne Cason Laing. Neither a writer nor a historian, Laing was vague about what was in the uncataloged pile of papers that she identified as "the Clum manuscript," which she turned over to Boyer in 1967. Over the years, her recollection would change about the source of the papers and which copies of the Cason-Ackerman manuscript were burned and which were kept. But neither Laing nor her siblings repudiated the idea that Josephine had "started a manuscript" with John Clum. In a 1983 affidavit, Laing made this clear: "My mother and Aunt were aware of the earlier 'Clum' manuscript covering the Tombstone years, and for that reason were willing to burn that portion of their manuscript at Mrs. Earp's request. My Aunt had written that portion."

When experts demanded that Glenn Boyer produce the Clum manuscript, he refused, which was taken as evidence of more chicanery on his part. But Boyer could not show anyone the Clum manuscript because it never existed as a "manuscript," i.e., a single document of some internal coherence such as the Cason manuscript. So what was in that "stack of papers almost a foot high"? Probably a precious heap of handwritten letters and notes relating to the Tombstone era, which Clum and others had gathered for Josephine, supplemented by Mabel's own research, her extensive correspondence in the 1950s as she went in and out of partnership with John Gilchriese, and then more letters, clippings, and research notes about the momentous discovery of the Mattie Blaylock story.

It has been a long time since anyone looked beneath the deceptive surface that Josephine herself created, never guessing that she would be followed by others with their own reasons to extend her trail of deceit. The advent of online message boards for the Earp faithful kept the idea of a conspiracy front and center, with extravagant claims that Boyer had perpetrated the greatest hoax since the Clifford Irving autobiography of Howard Hughes or the forged Hitler diaries. Anonymity of comments lent to the discussion the occasional raw sewage scent of anti-Semitism and homophobia.

The ad hominem attacks displaced the focus on history. The result was that in the six or so decades since Josephine's death, she has had no honest broker like Mabel Cason.

One wonders, what would Mabel do now?

UNTIL THE END of her life, Jeanne Cason Laing stayed in touch with Glenn Boyer, trading stories about hip replacements, hearing aids, and the endless round of doctor's appointments that consume much of octogenarian life. Their relationship survived a rough patch when her family demanded the return of the Cason manuscript, arguing that Ernest had intended a loan rather than a gift. Boyer reacted with his usual restraint. He threatened a lawsuit and unleashed his own accusations of a "criminal conspiracy" motivated by the greed of the younger Cason generation. The conflict was never resolved, and today, the manuscript sits in the Dodge City Historical

Society, with Laing's affidavit affixed on top, available only to those who have received Boyer's permission in advance.

Telling the truth about Mrs. Wyatt Earp was a responsibility that Laing felt she owed to Josephine, to the Cason family, and to history. She was bothered by the distortions that threatened to obscure the Aunt Josie she knew so well. "Eventually, no one will really know about the charming and aggravating little girl who ran away from a good home and into 'the wild west' to find the love of her life, but who never forgot that she was Josephine Marcus of San Francisco," she reflected.

On her eighty-fourth birthday, Glenn Boyer called Jeanne Cason Laing for the last time. "Are you having trouble with men chasing you?" he inquired, assuring her that he was still "blindingly handsome," but hadn't pulled off a seduction in months. Sorry to hear that she too was now using a walker, he blamed her sore back on her ample bosom. They chuckled together like old friends who accepted their losses as long as they were able to laugh, especially at themselves. There was not a speck of self-pity in either one of them.


AND NOW, ABOUT that curse.

I guess I should be worried. After all, this is the book that Josephine never wanted to be written. Hide your history, she demanded of herself—or at least control it. The story of the breathless beauty who stole handsome Wyatt away from his hapless wife would shock and alienate anyone who heard about it.

Wouldn't it?

Josephine never dared to believe that she would be forgiven. She cared too much about public judgment to take any chances. Time, however, was on her side. Today, we can see her as a passionate, resilient, unconventional woman, a myth builder for a bold new American era, with a remarkably modern understanding of media and celebrity. But she deserves far more than just forgiveness. She earned her recognition as an artist of adventure who emerged from one cocoon as an immigrant Polish Jew to mine the dynamic energy of the frontier. Even when the frontier mutated into a sterner, more constrained America, she fashioned new fables that fit the legend of Wyatt and Josephine Earp to the new century.

In that long road trip she took from Tombstone to Hollywood with Wyatt, she was not always the driver, but you can bet that she was never in the back seat. She was the one holding the map and laughing all the way.

And how ironic that it should fall to another Jewish woman to wonder if Josephine Sarah Marcus Earp would forgive me.

Or is there a statute of limitations on curses?

Either way, I've decided to believe that Josephine would not curse me for writing her back into American history.

And long may her story be told.

# | Acknowledgments

THANKS TO the ever curious Michele Slung for asking that fateful question: How did Wyatt Earp end up in a Jewish cemetery? The year was 2008, and I was thinking about whether I could write another biography.

My first book was *Sala's Gift*, an intensely personal journey through my mother's history and heroism during the Holocaust. She inspired my passion to uncover the stories of strong women and the circumstances that shape their lives. She and my father Sidney Kirschner are the greatest sources of strength and wisdom for me and for three generations, now including seven beautiful great-grandchildren: Hannah, Michelle, Joseph, Jordan, Nathan, Samuel, and Jacob.

For this book, I traveled with a powerful posse that assured me that yes, I could juggle a family and another book and a dean's responsibilities and ambitions.

Elisabeth, Caroline, and Peter taught me that parenting is surely the most joyous adventure of all, especially in the company of such a remarkable trio, now making their way in the world with heart, soul, and style.

Josephine never had a best friend. I do, and it is my good luck that Lorraine Shanley allowed the Earps to elbow their way into our Scrabble games and phone marathons. We are opposites only in being blond and brunette and a few other details; we are twins in all the ways that matter.

When I started writing, Jane and Bob Stine were not yet my *machatunim*. We hit the in-law jackpot when our families were joined officially by the marriage of our children. Jane's wit and wisdom informed my efforts at every turn.

My agent and friend, Flip Brophy, declared that I *would* finish this book, and so I have: when Flip believes in you, all things really do seem possible.

Gail Winston, my remarkable editor at HarperCollins, took a mess of ideas and events and characters and led me to the story. She and Maya Ziv and Joanne Pinsher have been wonderful colleagues and friends, and I thank the extended HarperCollins family for creativity and commitment to this book.

Macaulay Honors College is so much more than my day job; it is my daily inspiration. I am grateful to Chancellor Matthew Goldstein, who should win a prize for being my longest-running boss, to my dear colleagues, and to the Macaulay students and instructional technology fellows, especially my gang of hardworking and brilliant research assistants: Dan Blondell, Cecille Bernstein, Virginia Milieris, Savannah Gordon, Katherine Maller, Mary Williams, David Kane, Ayelet Parness, and Gregory Donovan. Thanks for teaching me so much.

I have tipped my hat to my many Earp history colleagues in the endnotes. I hope you will consider my inevitable mistakes and omissions to be honest ones. A special note of appreciation to Casey Tefertiller, Jeff Guinn, Eric Weider, Mark Ragsdale, and Roger Peterson for a myriad of professional courtesies, plus the fun of shared speculation, and more than one timely shout of "Danger, Will Robinson!" I am deeply indebted to my distinguished panel of first readers: Lynn Bailey, Anne Collier, Bruce Dinges, Karen Franklin, Ava Kahn, and Bob Palmquist. Your knowledge saved me from many an infelicity. To the one and only Glenn Boyer and Jane Candia Coleman, thank you both for welcoming a New Yorker and an "acaDU-Mic" to Planet Earp, and for opening those doors to the past to which only you have the keys.

Dr. Walter Cason played an indispensable role as my one true link to Josephine and Mabel Cason. I am grateful for his generosity and grace.

And finally, to my husband, Harold Weinberg. Ever loving and patient, you have made my choices your priorities. Had medicine not been your calling, you could have been a fine editor. I marvel at the twist of good fortune that brought us together in 1968 to build a charmed life together. I never hesitated to believe in the triumph of Josephine's nearly fifty-year marriage of love, laughter, and loyalty, because I know we too are on that journey.

# | Earpnotes

Lucky me, I had the benefit of interviews and insights from a group of people who don't know each other—and in some cases, don't like or trust each other. That's Planet Earp.

I would like to acknowledge conversations and contributions from Fred Agree, Allan Barra, Lynn Bailey, Mark Boardman, Glenn Boyer, Walter Cason, Robert Chandler, Jane Candia Coleman, Anne Collier, Burt Devere, Dorothy Devere, Dino DiConcini, Hasia Diner, Bruce Dinges, Scott Dyke, Mark Dworkin, Marge Elliott, Stephen Elliott, Karen Franklin, Leslie Fried, Tom Gaumer, Murdock Gilchriese, Gary Greene, Jeff Guinn, Rachel Tarlow Gul, George Laughead, Denise Lundin, Roger Lustig, Ray Madzia, Felton Macartney, Paula Marks Mitchell, Carol Mitchell, Jeff Morey, Christine Rhodes, Harriet Rochlin, Ava Kahn, Marguerite La Riviere, Bev Mulkins, Kevin Mulkins, Robert Palmquist, Roger Peterson, Bob Pugh, Mark Ragsdale, Gary Roberts, Christine Rhodes, Alice Rogoff, John Rose, Nona Safra, Laura Samuelson, William Shillingberg, Casey Tefertiller, Suzanne Westaway, Jeff Wheat, Steve Weiner, and Eric Weider.

The single most important source for the true story of Josephine Marcus Earp is the Cason manuscript of her memoir, based on interviews with her and written by Mabel Earp Cason and Vinnolia Earp Ackerman. I have mined the manuscript but cross-checked its assertions with many other sources. Unless otherwise indicated, all references are from this manuscript.

With some notable exceptions, Josephine's letters are still mostly in the hands of private collectors, and I am grateful to those who have shared their treasures with me. In particular, I am deeply indebted to Mark and Lauri Ragsdale and their children Wyatt and Isabella for their warm hospitality. All quotes from Josephine's letters not otherwise noted are from the Ragsdale Collection, which also included some of Flood's original notes and diagrams from his interviews with Wyatt Earp, as well as photographs, clippings, and documents.

Glenn Boyer granted permission to use his papers at the University of Arizona and Dodge City, and provided many other documents and photographs, and recordings of the extensive interviews that he did with the Cason and extended Marcus families; many of these are discussed here for the first time. Casey Tefertiller provided access to his recorded interviews with the Cason and Welsh families, and with Frank Waters and his wife. Roger Peterson shared his recorded interviews with William S. Hart, Jr., Jeanne Laing, and the Marcus family, as well as some unpublished articles. Suzanne Westaway provided access to the personal papers of Edna Lehnhardt Stoddart.

First stop for any Earp researcher is the remarkable Stuart Lake Collection at the Huntington Library. The papers of the Houghton Mifflin Company, the preeminent publisher of early western-themed books, reside at the Houghton Library at Harvard, and include important correspondence with Stuart Lake, William MacLeod Raine, Frederick Bechdolt, and William Breakenridge.

Other important archival sources include the Zaff-Behan divorce papers at the Arizona Historical Society Collection in Tucson, Behan's papers at the University of Arizona Special Collections, the recorded interview with Hildreth Halliwell at the University of Arizona Special Collections, and the letters of Louisa Earp at the Ford County Historical Society, with permission from Glenn Boyer. There is a treasure trove of Tombstone real estate, voting, and legal records at the Cochise County Courthouse and County Recorders Office in Bisbee, Arizona. The records of the first synagogues in San Francisco are in the Western Jewish History Center at the Judah L. Magnes Museum at the University of California, Berkeley. The records of the Marcus family burial plots are in the Hills of Eternity cemetery in Colma, California. The archives of the Alaska Commercial Company reside in Green Library at Stanford University. I spent an unforgettable week in Alaska, working in Nome's Carrie McLain Memorial Museum and the Kegoayah Kozga Library, and the University of Anchorage Special Collections.

I traveled back to the early days of Nome in the company of two fine writers, popular in their own day and hardly known at all today. Rex Beach's best seller *The Spoilers* deserves yet another generation of readers, or perhaps another movie version to add to the five that have already been made, beginning with a silent film in 1914 (preceding *The Birth of a Nation*) and ending with a 1942 version starring John Wayne and Marlene Dietrich. I also feel privileged to have stumbled upon the work of actor, writer, and activist Elizabeth Robins. Her time in Nome in search of her brother Raymond inspired two novels and several short stories. Her journal was edited by Victoria J. Moessner and Joanne E. Gates and published as *The Alaska-Klondike Diary of Elizabeth Robins, 1900*. The original is in the Fales Library and Special Collections of New York University.

The development of the San Francisco Jewish community is an important backdrop for Josephine's life. For the Marcus family, the tale began in Europe and in the Prussian province of Posen (sometimes known as Posnan); I found Hasia Diner's *A Time for Gathering: The Second Migration, 1820–1880* an invaluable introduction, which led me to Isaac Benjamin and his fascinating account of Jewish life in America right after the Civil War, *Three Years in America: 1859–1862*. Benjamin did have a prejudice against the Emanu-El crowd, as Ava Kahn notes in her indispensable *Jewish Voices of the California Gold Rush*.

For excellent histories of Tombstone, see William Shillingberg, *Tombstone, A.T.: A History of Early Mining, Milling, and Mayhem*, and Lynn Bailey's *Tombstone, Arizona: "Too Tough to Die"; The Rise, Fall, and Resurrection of a Silver Camp, 1878 to 1990*. All Tombstone historians are indebted to Lynn Bailey and Don Chaput for *Cochise County Stalwarts: A Who's Who of the Territorial Years*.

And then there is Wyatt Earp and the O.K. Corral. Most roads begin with Stuart Lake's *Frontier Marshal*, Walter Noble Burns's *Tombstone: An Iliad of the Southwest*, and William Breakenridge's *Helldorado: Bringing the Law to the Mesquite*. With a groaning bookshelf of modern Wyatt Earp biographies to choose from, the discriminating reader will start with Casey Tefertiller's *Wyatt Earp: The Life behind the Legend*. Jeff Guinn's *Last Gunfight* is the most definitive (and most enjoyable) book about Wyatt Earp in decades. Other important sources include Allan Barra, *Inventing Wyatt Earp: His Life and Many Legends*; Don Chaput, *The Earp Papers: In a Brother's Image*; William Shillingberg, *Dodge City: The Early Years, 1872–1886*; and Gary Roberts, *Doc Holliday: The Life and Legend*.

Roger Jay filled in some critical gaps—and documented the existence of Wyatt's first common-law relationship, with Sally Earp—in *Wyatt Earp's Lost Year* and *Reign of the Rough-Scuff, Law and Lucre in Wichita.* Carol Mitchell wrote one of the earliest and most important reconsiderations of Josephine in her article "Lady Sadie." I am indebted to Anne Collier for new documentation on Josephine's early days, and to Bob Palmquist for his encyclopedic knowledge of all things Tombstone.

For the essential trio of Tombstone contemporary observers, see George Parsons's original diaries at the Arizona Historical Society, Tucson. A portion of the diaries has been edited by Lynn R. Bailey in *A Tenderfoot in Tombstone: The Private Journal of George Whitwell Parsons; The Turbulent Years, 1880–82.* Endicott Peabody's observations on Tombstone are available in *A Church for Helldorado: The 1882 Tombstone Diary of Endicott Peabody and the Building of St. Paul's Episcopal Church.* Of the three observers, only Clara Spalding Brown wrote for the public; see Lynn R. Bailey, *Tombstone from a Woman's Point of View: The Correspondence of Clara Spalding Brown, July 7, 1880, to November 14, 1882.*

I am grateful to these people and places, especially to the heroic librarians and archivists who are a writer's best friend.

For additional sources, see ladyattheokcorral.com.

# | Notes

PROLOGUE: IN WHICH I LAND ON PLANET EARP

4    *As Nora Ephron would memorably say of herself:* Ephron was interviewed by Abigail Pogrebin in *Stars of David: Prominent Jews Talk about Being Jewish* (New York: Broadway Books, 2005), 168–73.

4    *her silhouette still amazed Grace Welsh Spolidoro:* Interview by Casey Tefertiller, November 8, 1994. Grace compared Josephine to Dolly Parton.

5    *June 2, 1860:* Josephine's birth records are missing, as is often the case with pre–Civil War birth records in New York City. Josephine herself wanted confirmation of the date: on January 18, 1942, she sent a letter (typed by John Flood) to the New York City Department of Health to ask for records of her birth year, emphasizing that "Josephine Marcus was born in New York City; I am positive."

6    See Carolyn G. Heilbrun, *Writing a Woman's Life* (New York: Norton, 1988).

8    *"Uncle Wyatt's old hang out":* From the unpublished diary of Edna Lehnhardt Cowing Stoddard Siegriest, November 27, 1954.

CHAPTER 1: A JEWISH GIRL IN TOMBSTONE

13    *just before Christmas, 1880:* Josephine refers in the Cason manuscript to her arrival as "holiday" time, and told Mabel Cason that her arrival coincided with the death of John Clum's wife. Mary Ware Clum died in Tombstone on December 18, 1880.

14    *in the province of Posen:* Hyman, the son of Abraham and Chana Marcus and apprenticed to his father's profession as a baker, was likely from Nakel, Posen. Note that spellings are

notoriously tricky for nineteenth-century European/American names. Posen is sometimes known as Posnan. Prussia is sometimes misidentified as Russia. Hyman Marcus is sometimes known as Herman or Henry or Carl-Hyman Marcuse. In the entry for the Marcus family plot in the Hills of Eternity, and on his tombstone, he is Hyman Marcus. Although Sophia is identified as "Sophia Lewis" in most records, her daughter Rebecca (Josephine's half-sister) was identified as "Levy" in her marriage announcement in the *Daily Alta California* on May 4, 1870. For a fuller picture of the Jewish migration from Posen and the internal prejudices, see Hasia Diner, *A Time for Gathering, the Second Migration, 1820–1880* (Baltimore: Johns Hopkins University Press, 1992), and Ava F. Kahn and Ellen Eisenberg, "Western Reality: Jewish Diversity During the 'German' Period" in *American Jewish History*, Volume 92, Number 4, December 2004, pp. 455–479.

16    *a remarkable Jewish success story:* The Jewish population was between 7 and 8 percent of San Francisco's total population by 1870, while Jews made up an estimated 2 to 5 percent of New York City's population. Moses Rischin and John Livingston, *Jews of the American West* (Detroit: Wayne State University Press, 1991), 34.

19    *Rebecca married Aaron Wiener:* They were married by Reverend H. A. Henry, the rabbi at Sherith Israel, one of two synagogues that had been established during the gold rush. The Marcus family synagogue followed the traditions of the Jews from Posen, while Temple Emanu-El catered mostly to the wealthier and more assimilated German Jews. Aside from the clothing, Josephine had no recollection of the wedding, and Mabel Cason later admitted that she made up the accurate but generic description of the Jewish wedding ritual. In fact, Josephine never revealed her sisters' married names in her memoir.

20    *a less than perfect German accent signaled "second class":* German/Polish tensions figure prominently and poignantly in Harriet Lane Levy's memoir *920 O'Farrell Street* (Garden City, N.Y.: Doubleday, 1947). As had Josephine's, Levy's parents had been born in Posen. Unlike Josephine's father, her father had been commercially successful. But that made no difference: she too carried the burden of inferiority. "On the social counter the price tag 'polack' confessed second class," Levy recalled. "Upon this basis of discrimination everybody agreed and acted." These were simply the facts of life: she and her classmates accepted their place below German-Jewish girls "as the denominator takes its stand under the horizontal line." Like Josephine, Harriet pretended she was German. However, she went on to graduate from Berkeley and moved to Paris with Alice B. Toklas, where she became an intimate of the Gertrude Stein salon.

21    *David Belasco or Gertrude Stein:* Neither Belasco nor Stein retained any traditional Jewish ties. Belasco preferred the dramatic fiction of a childhood in a Roman Catholic monastery to his father's history as former London Jewish clown. Stein, who once proclaimed herself the most famous Jew in the world, left Oakland for Paris, where she would eventually accept the patronage of Nazi protectors to save her life and her art collection.

23    *the popular actress Pauline Markham:* Pauline Markham's real name was Margaret Hall. *Daily Alta California*, December 28, 1884.

26    *"Miss Pauline Markham of the Pinafore Troupe":* *Daily Evening Bulletin*, January 28, 1880.

26 *her older sister and brother-in-law dispatched a family friend:* Josephine identifies this man as Jacob Marks, a friend of Aaron Wiener (sometimes spelled Weiner).

34 *one of the great migrations of American history:* Between 1840 and 1870 about 350,000 people made the trip west, most of them in search of land and minerals. In *Women's Diaries of the Westward Journey* (New York: Schocken, 1982), Lillian Schlissel observed that in their private writings, men tended to see it as a "mythic" adventure, while the women were more likely to emphasize the family sacrifices of the journey.

34 *threatened to whip if their parents did not:* "Diary Written by Wife of Dr. J.A. Rousseau," *San Bernardino Museum Association Quarterly* 6 (Winter 1968). Quoted in Donald Chaput, *The Earp Papers: In a Brother's Image* (Encampment, Wyo.: Affiliated Writers of America, 1994), 14.

36 *Sally Haspel:* Roger Jay was the first to establish a link between Wyatt Earp and Sally Haspel, also known as Sallie Haskell, in "Wyatt Earp's Lost Year," *Wild West* 16, no. 2 (August 2003): 46. http://www.historynet.com/wyatt-earps-lost-year.htm.

37 *Their friendship was sealed:* The relationship between Wyatt Earp and Doc Holliday has been discussed in the comprehensive biography by Gary Roberts, *Doc Holliday: The Life and Legend* (Hoboken, N.J: John Wiley & Sons, 2006), and in Karen Tanner and Robert K. DeArment, *Doc Holliday: A Family Portrait* (Norman: University of Oklahoma Press, 1998). For another point of view, see Andrew Isenberg, "The Code of the West: Sexuality, Homosociality, and Wyatt Earp," *Western Historical Quarterly* 40, no. 2 (2009): 139, which argues that Josephine was a "heteronormative" replacement for Holliday.

37 *until she met Wyatt, probably in Fort Griffin:* Other accounts have them meeting in Fort Dodge or Fort Scott, as early as 1877 or 1878. Although Stuart Lake declared that during Wyatt's years in Kansas he "steered pretty clear of entangling alliances," Wyatt had a girl in every camp. Glenn Boyer discusses these years in *Suppressed Murder of Wyatt Earp* (San Antonio, Tex.: Naylor, 1967), 112.

38 *common-law marriages were easy to create:* Common-law marriage originated in pre-Reformation England and ended in 1753 when Parliament passed Lord Hardwicke's Act, requiring all marriage ceremonies to be performed by officials of the Church of England. Nonetheless, common-law marriage traveled across the Atlantic to the Americas. Arizona banned common-law marriages in 1913. See Cynthia Prescott, " 'Why She Didn't Marry Him': Love, Power, and Marital Choices on the Far Western Frontier," *Western Historical Quarterly* 38, no. 1 (Spring 2007): 25–46. The famous Lotta Crabtree case is even more directly relevant to Wyatt and Josephine; see David Dempsey and Raymond P. Baldwin, *The Triumphs and Trials of Lotta Crabtree* (New York: Morrow, 1968).

38 *left San Bernardino as soon as he sent for her:* Louisa appeared to have been deeply in love with Morgan, but chafed at the famous Earp solidarity, which apparently extended to having her mail opened by Morgan's brothers. She directed her sister to write to her as Louisa Houston, since "any letter that is not directed to this name is allways opened by my husbands brothers."

40  *lesbian couple embracing in public:* George Parsons's entry of July 16, 1880, describes this Tomb-
    stone scene: "I have seen hard cases before in a frontier oil town where but one or two women
    were thought respectable but have never come across several such cases as are here. It would
    be impossible to speak here of some or one form of depravity I am sorry to know of—for bad
    as one can be and low as woman can fall—there is one form of sin here fortunately confined to
    two persons which would I almost believe bring a blush of shame to a prostitute's cheek."

41  *"Everything was nice if you had money, and we didn't so it wasn't":* For two essential articles on
    the controversy surrounding Allie Earp's recollections, as filtered through Frank Waters,
    see Gary Roberts, "Allie's Story: Mrs. Virgil Earp and the *Tombstone Travesty*," http://
    home.earthlink.net/~knuthcol/Travesty/AlliesStory1source.ht, and Casey Tefertiller,
    "What Was Not in *Tombstone Travesty*," http://home.earthlink.net/~knuthcol/Travesty/
    notintravestysource.htm.

42  *Earp's stepdaughter had married:* Hattie Catchim Earp, Bessie's daughter, is often (errone-
    ously) thought to have run off with one of the Tombstone cowboys. See Anne Collier,
    "Harriet 'Hattie' Catchim: A Controversial Earp Family Member," *Western Outlaw-
    Lawman History Association (WOLA) Journal* 16, no. 2 (Summer 2007).

46  *Johnny Behan with another woman:* Some accounts have Josephine returning from a trip to
    San Francisco and catching Johnny in bed with another woman. Mabel Cason wrote on
    May 20, 1959, that Josephine "told us about her time in Tombstone, how she had gone there
    with a theatrical troupe playing Pinafore, that she had met John Behan there and went back
    later with the understanding that they were to be married, kept house for him, looked after
    his 10 year old son Albert. She built a house for them with money that her father had sent
    her and some derived from the sale of her diamonds. Finally [Behan] began running around
    with a married woman and neglecting her—and she met Wyatt Earp." Original letter is in
    the Arizona Historical Society, Tucson.

46  *a likely candidate was Emma Dunbar:* Emma Dunbar's letters to John Behan (Special Col-
    lections, University of Arizona) document that they had an affair in Tombstone, though
    the timing is not conclusive. Emma and Johnny corresponded for years afterward; when
    she heard that he was headed for Alaska, she warned him that he was not up to the rigors
    of the frozen north: "My dear boy what in heavens name would you with your dainty ways
    and white hands do in a country where men who have roughed it all their lives die. . . . You
    have not the strength or endurance, and why should you go?"

46-47 *disturbing signs that Johnny had contracted syphilis:* Although the date when Behan con-
    tracted syphilis is unknown, John Gilchriese believed that he was infected in Tombstone.
    It could have been as late as the mid-1880s. Lynn Bailey reports Behan's trips to the West
    Coast for treatment are documented in newspapers of the late 1880s and '90s. His disease
    reached the tertiary stage by the early 1890s, and eventually contributed to his early death.

47  *Tombstone's sex trade was regulated:* For discussions of prostitution in Tombstone, see Ben
    Traywick, *Behind the Red Lights* (Tombstone, Ariz.: Red Marie's, 1993); Anne M. Butler,
    *Daughters of Joy, Sisters of Misery: Prostitutes in the American West, 1865–90* (Urbana: Uni-

versity of Illinois Press, 1985); and Anne Seagraves, *Soiled Doves: Prostitution in the Early West* (Hayden, Idaho: Wesanne, 1994).

49   *"the key to the whole yarn of Tombstone":* Stuart Lake to Ira Rich Kent, February 13, 1930, Houghton Mifflin Collection, Houghton Library, Harvard University.

53   *"very grave results will follow":* Quoted in Paula M. Marks, *And Die in the West: The Story of the O.K. Corral Gunfight* (New York: Morrow, 1989), 174–75.

55   *Who shot first?* The arguments rage on. For the best recent comprehensive account, see Jeff Guinn's *The Last Gunfight: The Real Story of the Shootout at the O.K. Corral and How It Changed the American West* (New York: Simon & Schuster, 2011).

59   *Spicer issued a strong verdict:* The complete inquest records were recently discovered in Cochise County and are online at http://azmemory.lib.az.us/cdm/search/collection/ ccolch. For a well-documented summary of some of the key evidence presented at the hearing, see Jeff Morey, " 'Blaze Away!' Doc Holliday's Role in the West's Most Famous Gunfight," http://home.earthlink.net/~knuthcol/Itemsofinterest4/blazeawaysource.htm.

60   *the Grand Hotel across the street:* There is some evidence that Mattie stayed behind when the rest of the family moved to the hotel. See Marks, *And Die in the West*, 241.

61   *a comedy at Schieffelin Hall:* It is baffling, but provocative, that in the manuscript of Flood's biography of Wyatt Earp in the Ford County Historical Society, he has Wyatt accompany Morgan to a Pauline Markham production of *Pinafore*, not *Stolen Kisses*. The Markham troupe (including Josephine) had indeed performed *Pinafore* in Tombstone, but that was in 1879, more than two years before Morgan's death. "Here was Pauline Markham, and Pinafore," Flood has Morgan appealing to Wyatt. "There would be nothing like it for another year." I would suggest that this was an inside joke planted by Josephine, except that nothing about Morgan's death was funny.

63   *"saves the Earp family from annihilation":* Wyatt later made it clear that he did not believe that Pete Spence was responsible for Morgan's assassination, though Josephine reiterated this accusation in the Cason manuscript. It is possible that Maria Spence took advantage of the situation to rid herself of her abusive husband. Spence (whose real name was Elliott Larkin Ferguson) eventually divorced Maria and married the widow of Phin Clanton. See Lynn R. Bailey and Donald Chaput, *Cochise County Stalwarts: A Who's Who of the Territorial Years* (Tucson, Ariz.: Westernlore Press, 2000).

63   *Josephine's departure went unnoticed:* The exact date of Josephine's departure and her method of transportation is unknown. One clue is that her friend Annie Lewellen sent a money order to Josephine's mother in San Francisco on March 22, 1882, presumably on Josephine's behalf or perhaps in repayment of a debt to Josephine.

## CHAPTER 2: THE FOURTH MRS. EARP

70   *There was money to be divided:* Wells Fargo historian Robert Chandler has analyzed Wyatt's payments during this time. See "Under Cover for Wells Fargo: A Review Essay," *Journal of Arizona History* 31 (Spring 2000): 83-96; and "Wells Fargo and the Earp Brothers: The Cash Books Talk," *California Historical Quarterly*, no. 78 (Summer 2009): 5-13.

70  *Wyatt stayed with his friend Henry Jaffa:* See Mark Dworkin, "Henry Jaffa and Wyatt Earp: Wyatt Earp's Jewish Connection; A Portrait of Henry Jaffa, Albuquerque's First Mayor," *New Mexico Jewish Historical Society* 19, no. 4 (December 2005): 25-37.

70  *"a damn Jew boy":* This incident was covered in the local newspapers at the time. For a discussion about the so-called Ortero letter, the source of Doc's "Jew boy" comments, see Chuck Hornung and Dr. Gary L. Roberts, "The Split: Did Doc and Wyatt Split Because of a Racial Slur?" *True West*, December 2001, 58-61.

71  *George W. Crummy, an influential saloonkeeper:* Gary Roberts identified Crummy in *Doc Holliday*, 308. At the time, Wyatt was vague about this connection, and he later told Flood, "Governor Pitkin was an old man, about 70 years of age. . . . One man only could influence him, a man by the name of Crummy. Now Governor Pitkin was associated with Crummy (owners of a gambling house) in running enterprises in San Juan district in CO." Flood's notes, interview with Wyatt Earp, September 15, 1926, Ragsdale Collection.

72  *the luxurious Tabor Hotel:* Josephine may have been confusing the Grand Hotel in Leadville with the Windsor Hotel in Denver; both were owned by Tabor. See Duane A. Smith, *Horace Tabor: His Life and the Legend* (Boulder: University Press of Colorado, 1989), 260.

73  *"glowing account":* Alas, this article has not yet come to light.

74  *house on Telegraph Avenue:* Glenn Boyer, taped interview with Marjorie Macartney (Josephine's grandniece), May 27, 1984. As of this writing, the Oakland address at 2703 Telegraph is the forlorn site of an abandoned wig store.

74  *registered at the Washington Hotel in Galveston.* In December 1883, Josephine was receiving mail in Galveston as both Mrs. Josephine Earp and Mrs. Sadie Earp. See *Galveston Daily News*, December 1 and 23, 1883.

74  *had the bad luck to visit Globe:* In *Suppressed Murder of Wyatt Earp*, Glenn Boyer attributes this anecdote to John Gilchriese. In *I Married Wyatt Earp: The Recollections of Josephine Sarah Marcus Earp* (Tucson: University of Arizona Press, 1976), 129, Boyer also claims that Wyatt and Johnny Behan had a fistfight in Globe, and that Wyatt "knocked him out cold."

77  *San Diego native William J. Hunsaker:* Hunsaker's father was sheriff of San Diego County. Hunsaker eventually became district attorney, mayor of San Diego, and president of the Los Angeles Bar Association. He would remain the Earps' lawyer until his death. See Richard F. Pourade, *The History of San Diego* (San Diego, Calif.: Union-Tribune, 1960), chapter 13, http://www.sandiegohistory.org/books/pourade/glory/glorychapter13.htm.

82  *Josephine returned less often to San Francisco:* Garner Palenske and Ben Traywick traced Josephine's steamship voyages from San Diego to San Francisco in *Wyatt Earp in San Diego: Life after Tombstone* (Santa Ana, Calif.: Graphic, 2011).

82  *Wyatt's absences from San Diego grew more frequent:* For all of Josephine's nostalgia for San Diego, there is speculation that Wyatt and Josephine separated briefly during these years. Roger Peterson has pointed out that there is a "Mrs. Earp" listed separately in the city directory, which might lead to the conclusion that Mattie had followed Wyatt there. This would be a remarkable act by the ordinarily passive Mattie. The less dramatic, but more

likely, explanation is that Wyatt put the apartment in Josephine's name, as he once did for Mattie in Tombstone.

84 *the sordid details of Mattie's death:* See Celia Earp Inquest, Arizona Department of Library, Archives Division, Pinal County Inquests, /filmfile 88.6.1. Quoted in Casey Tefertiller, *Wyatt Earp: The Life Behind the Legend* (New York: J. Wiley, 1997), 280.

85 *There is no evidence that Baldwin ever owned a yacht, though his daughter did:* On July 26, 1916, Clara Baldwin Stocker's yacht was boarded by English military who were searching for German submarines. On the same journey, Stocker was nearly thrown overboard by the rough seas and lost an expensive diamond. *New York Times,* July 27, 1916.

86 *"He Is a Dude Now":* *Denver Republican,* March 14, 1893.

87 *young historian Frederick Jackson Turner:* Turner's thesis has been reviewed, refuted, defended, and revised many times by historians who take him to the woodshed for a myopic view of America, one without women and Native Americans, for instance. That said, the delicious irony of Josephine and Wyatt in close proximity to Turner's star performance should not be overlooked. See Frederick Jackson Turner, "Social Forces in American History," presidential address before the American Historical Association, *American Historical Review* 16 (1910): 217–33; and John Mack Faragher, ed., *Rereading Frederick Jackson Turner: The Significance of the Frontier in American History and Other Essays* (New York: Holt, 1994).

## CHAPTER 3: THE GREATEST MINING CAMP THE WORLD HAS EVER KNOWN

89 *the desert outside of Yuma, Arizona:* See Mark Dworkin, "Wyatt Earp's Yuman and Cibola Sojourns," *Western-Outlaw Lawman History Association Journal* 14, no. 1 (Spring 2005).

91 *"Fitzsimmons Was Robbed!":* See *San Francisco Call,* December 3, 1897. The Sharkey-Fitzsimmons fight was held at the Mechanics' Pavilion, located at Larkin and Grove Streets, San Francisco, the site of the present Civic Auditorium. The controversy was well covered on the East Coast: see "Sharkey Gets $8500," *New York Times,* December 18, 1896. Even now, there is still disagreement among experts as to whether Wyatt made a good call, a bad call, or a corrupt call. For a nuanced reconsideration, see Gary Roberts in *The West,* November 1971, pp. 10–52.

94 *hunger, fever, dysentery, and scurvy:* See E. Hazard Wells and Randall M. Dodd, *Magnificence and Misery: A Firsthand Account of the 1897 Klondike Gold Rush* (Garden City, N.Y.: Doubleday, 1984).

94 *Josephine discovered that she was pregnant:* Although Josephine was more forthcoming about her wedding on Lucky Baldwin's yacht than she was about her pregnancies, there is strong evidence for this. The 1897 miscarriage was apparently "family lore," according to Wyatt's niece and nephew, Estelle and Bill Miller, and Josephine's grandniece, Alice Cohen Greenberg, who were interviewed extensively by Glenn Boyer. Jeanne Cason Laing also believed that Josephine had suffered a miscarriage that year, according to Casey Tefertiller.

94 *the stillborn child was a boy:* John Flood wrote to Edward Earp on June 10, 1952: "Also, you probably are aware that the children of Wyatt Earp died in infancy; I understood there were two boys." Assuming that Flood was correct, one likely interpretation would be that one

male child died with Aurilla Earp, and the other was Josephine's stillborn child, although it is possible that both sons could have been Josephine's lost babies. Ragsdale Collection.

97   *Josephine found a one-room cabin:* This may have been the cabin of Rex Beach. His papers are housed in part at his alma mater, Rollins College. Photographs of his cabin at Rampart are in the collection of Candy Waugaman.

100  *1,500 dozen precious eggs:* Josephine's sad story of the Egg Man is verified by a contemporary letter written from Rampart on November 1898: "Last night a man who bought 1500 *dozen* eggs packed in lard to open a restaurant was found unconscious—a suicide. His eggs were too ancient for use and he could not sell them so he grew despondent." Herbert Heller Papers, Lynn Smith Correspondence and Letters, University of Alaska, Fairbanks. See also Jack London's story with a similar plot, "A Thousand Dozen."

100  *the gateway to the Yukon:* The trip was the brainstorm of Erasmus Brainerd, a journalist and later head of the Seattle Chamber of Commerce, who believed that Alaska could be critical to the future of Seattle. He brought with him John H. McGraw, the former governor of Washington, and E. M. Carr, a lawyer and brigadier general in the Washington State militia. For histories of the relationship between Seattle and the Klondike gold rush, see Richard C. Berner, *Seattle, 1900–1920: From Boom-town Urban Turbulence, to Restoration* (Pullman: Washington State University Press, 1991); and Paula Mitchell Marks, *Precious Dust: The American Gold Rush Era, 1848–1900* (New York: W. Morrow, 1994).

108  *News of the first Nome strikes:* See T. H. Carlson, "Discovery of Gold at Nome, Alaska," *Pacific Historical Review* 15, no. 3 (September 1946): 163–175.

109  *"from sea-beach to sky-line the landscape was staked":* E. S. Harrison, *Nome and Seward Peninsula: A Book of Information about Northwestern Alaska* (Seattle: E. S. Harrison, 1905). For a review of the relevant mining laws, see Carl Mayer, "Mining Law: Historical Origins of the Discovery Rule," *University of Chicago Law Review* 53, no. 2 (Spring 1986): 624–653.

112  *Nome was struggling to become a habitable city:* See L. H. French, "The Beach at Cape Nome," in *Nome Nuggets: Some of the Experiences of a Party of Gold Seekers in Northwestern Alaska in 1900* (New York: Montross, Clarke & Emmons, 1901), 34–44.

114  *Some 1,000 miners were broke and homeless in Nome:* There were specific congressional appropriations to subsidize up to $75 for a return ticket home. Samuel C. Dunham, *The Yukon and Nome Gold Regions*, vol. 5, Bulletin No. 219, July 1900.

116  *Reed's Blizzard Defier Face Protector:* For the comprehensive story of Nome, as well as more on the relationship between Seattle and Nome, see Terrence Cole and Jim Walsh, *Nome: City of the Golden Beaches* (Anchorage: Alaska Geographic Society, 1984).

117  *Wyatt was "making money":* San Francisco Examiner, November 13, 1899. Gambling was technically illegal in Seattle in 1899, but a more accurate description would be that it was subjectively regulated. For a detailed account of Wyatt's brief success with the Union Club, see Pamela Potter, "Wyatt Earp in Seattle," *Wild West*, October 2007.

119  *Josephine and Wyatt left from Seattle on the* Alliance: The *Seattle Post-Intelligencer* of May 20, 1900, reported, "Wyatt Earp, the well known sporting man of Seattle and San Francisco, took passage on the *Alliance* for Nome where he operated in business and mining affairs last

year. He is accompanied by Mrs. Earp." The 1900 census was executed aboard the *Alliance* on June 14. The timing is important, because it establishes the Earps' whereabouts when Warren Earp was killed.

119 *Josephine entertained herself with gambling:* According to Grace Spolidoro, Wyatt complained to his friend Charlie Welsh that Josephine gambled heavily on the boat trips to Alaska.

120 *Alice with her husband Isidore:* In the Cason manuscript, Josephine mistakenly recalls her Alaskan guest as her sister Henrietta's daughter Edna, who was still a child. It was her sister Rebecca's daughter Alice who came to Nome.

120 *to take stock of dramatic changes:* This is a composite description drawn from accounts previously cited from L. H. French, Elizabeth Robbins, George Parsons, and Rex Beach, as well as a contemporary diary by Avaloo Boyd, "Alaska, I Love You," in Nome's Carrie M. McLain Memorial Museum.

122 *"and you have Nome as it was":* Quoted in Cole and Walsh, *Nome*, 59.

124 *Fitzsimmons would be fighting in Nome in advance of another bout with Sharkey: Nome Chronicle,* September 3, 1900.

127 *ordered Wyatt to get rid of the women:* From Roger Peterson's interview with Alice "Peggy" Greenberg, Josephine's grandniece, October 13, 1981, Roger S. Peterson Collection.

128 *a second vendetta to avenge Warren:* Jeff Morey conclusively demonstrated the impossibility of the Earps being in Arizona in "The Curious Vendetta of Glenn G. Boyer," *Quarterly of the National Association for Outlaw and Lawman History (NOLA)* 18, no. 4 (October–December 1994).

128 *Clum quickly revolutionized mail service in Nome:* After the personal and often violent attacks he endured in Tombstone, Clum must have treasured the appreciative public in Nome: "General Clum is a man of ripe experience, broad minded and practical," editorialized the *Nome Daily News*. "A frontiersman himself, he knows what to do and how to act in meeting emergencies." See Fred Lockley, *History of the First Free Delivery of Mail in Alaska at Nome, Alaska, in 1900* (Seattle: Shorey Book Store, 1966).

132 *Nome accounted for almost half of the total gold output:* Alfred H. Brooks, George B. Richardson, and Arthur J. Collier, *A Reconnaissance of the Cape Nome and Adjacent Gold Fields of Seward Peninsula, Alaska, in 1900* (Washington, D.C.: U.S. Government Printing Office, 1901).

133 *"that I wear as a clip":* Alice "Peggy" Greenberg, interview by Roger S. Peterson, October 13, 1981, Roger S. Peterson Collection.

133 *The mining claims that Wyatt had filed in Josephine's name:* Interestingly, Wyatt was not too famous to have his name misspelled on the documents for the sale of the Dexter or the transfer of the mining claims, which record his name as "Erp."

CHAPTER 4: WAITING FOR WYATT

136 *They outfitted themselves for the desert in Los Angeles:* See Carl B. Glasscock, *Gold in Them Hills: The Story of the West's Last Wild Mining Days* (Indianapolis: Bobbs-Merrill, 1932),

one of the best contemporary accounts on life in the boomtowns. Glasscock was subsequently the biographer of Lucky Baldwin. Josephine and Mabel Cason were familiar with his books. See also Jeffrey Kintop and Guy Louis Rocha, *The Earps' Last Frontier: Wyatt and Virgil Earp in the Nevada Mining Camps, 1902–1905* (Reno, Nev.: Great Basin Press, 1989).

137 *"[Wyatt Earp and his wife and A. Martin] are good citizens and we welcome them": Tonopah Bonanza,* February 1, 1902, quoted in Chaput, *Earp Papers,* 189.

138 *Rickard's reputation soared:* See Phillip I. Earl, "Tex Rickard—The Most Dynamic Fight Promoter in History," *Boxing Insider,* April 15, 2008, http://www.boxinginsider.com/history/tex-rickard-the-most-dynamic-fight-promoter-in-history/.

139 *Halliwell was usually quick to blame Josephine:* Halliwell, interview by Bill Oster and Al Turner, Colton, California, September 25, 1971, University of Arizona Special Collections.

139 *noted his obituary in the* Goldfield News*:* Jane was the daughter of Virgil and Jane Sysdam, who were married briefly in 1860 without parental consent. Allie sent Virgil's body to Jane for burial in Portland, Oregon. See Kintop and Rocha, *Earps' Last Frontier,* 40.

140 *only Wyatt was still alive*: Both Billy Claiborne and Ike Clanton were killed in shootouts; Claiborne in 1882 and Clanton in 1887.

142 *America's greatest urban catastrophe:* See Philip L. Fradkin, *The Great Earthquake and Firestorms of 1906: How San Francisco Nearly Destroyed Itself* (Berkeley: University of California Press, 2005).

143 *"Notorious Marshal Who Disqualified Fitzsimmons Arrested in Raid"*: See "Earp's Faro Plan Fails," *New York Times,* July 23, 1911.

143 *John Flood, a young engineering graduate:* According to Lynn Bailey, Flood worked in the Los Angeles offices of Seeley Wintersmith Mudd, an eminent geologist and founder of many companies, and father of Seeley G. Mudd, the prolific benefactor of American colleges, libraries, and science buildings.

146 *Wyatt's earnings declined:* One of these checks is in the Ragsdale Collection, drawn on the Central Bank of Oakland. Hildreth Halliwell considered Wyatt a "bitter man" at having to accept these handouts.

146 *Wyatt was paid a management fee:* Kirschner interview with Walter Cason, April 10, 2010; Boyer, interview with Louis Siegriest, 1983, Boyer Collection.

147 *The sight of Wyatt Earp . . . at a seder:* Melvin Shestack was the television producer who wrote about meeting Wyatt Earp and Henry Fonda in the *Forward,* July 1, 1994.

148 *condemn all those things she did when she was young:* The reverse story is told about Wyatt by the Welsh sisters. They too stayed with Josephine and Wyatt, but in their memory, it was Wyatt who "watched them like a hawk," and when they stayed too long at the beach, would say slowly in his deep voice, "Grace, time to come home." Casey Tefertiller, interview with Grace Welsh Spolidoro.

151 *"Cowboys" were in ample supply in the Los Angeles stockyards:* Raoul Walsh evokes this era in *Each Man in His Time: The Life Story of a Director* (New York: Farrar, Straus and Giroux, 1974), 75.

152 *"Tamed the baddies, huh?":* Walsh, 105.

153 *"those 'damn fool dudes,' as he called them":* Quoted in Joseph McBride, *Searching for John Ford: A Life* (New York: St. Martin's Press, 2001), 111–12.

153 *The "King of the Cowboys":* Aside from his accomplishments as an actor and stuntman, Tom Mix appears on the album cover of the Beatles' *Sgt Pepper's Lonely Hearts Club Band* (third row).

153 *"It haunted his mind":* Roger S. Peterson interview of William S. Hart Jr., April 6, 1983, Roger S. Peterson Collection.

155 *Hooker drafted a manuscript she showed to Wyatt:* The manuscript, "An Arizona Vendetta: The Truth About Wyatt Earp and Some Others," is in the Southwest Museum, Los Angeles.

156 *"Doc Holliday was about 5'10 ½", slender, good looking":* Flood notes, dated September 9, 1922, Ragsdale Collection. See ladyattheokcorral.com.

165 *Albert insisted that he was reporting the conversation accurately:* This is the version that Josephine tells in the Cason manuscript and in a letter she sent to Frank Lockwood. However, Wyatt Earp told Frederick Bechdolt that it was Albert's wife Julia Behan, who reported the conversation to Josephine, who then repeated it to Wyatt.

167 *There were lots of good married people there:* See David Dempsey and Raymond P. Baldwin, *The Triumphs and Trials of Lotta Crabtree* (New York: Morrow, 1968).

168 *"Old Wyatt Earp is still on deck":* William M. Raine to Ira Rich Kent, April 8, 1928, Houghton Library, Harvard University. Raine's article, "Helldorado: Stories of Arizona's Wild Old Days, When You Couldn't Keep a Bad Man Down," appeared in the magazine *Liberty* on July 16, 1927.

170 *Bat Masterson:* Lake may have known Masterson when he worked at the *Morning Telegraph* and Lake worked at the *New York Herald.* Kirschner interview with Anne Collier. See Collier, "Stuart N. Lake's Wyatt Earp and the Great Depression," B.A. thesis, University of La Verne, 2011.

171 *Josephine was forced to borrow money:* According to John Gilchriese and William Shillingberg, Josephine used the loan to pay gambling debts, and was barred from seeing Doheny again. Doheny was the model for Daniel Plainview in the film *There Will Be Blood.*

172 *interview Earp about six times:* Tefertiller, *Wyatt Earp*, 325.

173 *"straight as an arrow":* Roger S. Peterson, interview with Alice "Peggy" and Alvin Greenberg, October 13, 1981, Roger S. Peterson Collection.

174 *Adela Rogers St. John declared that she would never forget seeing Wyatt:* Quoted in Stephens, pp. 231–35. The original article was published in the *American Weekly,* May 22, 1960.

174 *The brothers had not been particularly close:* In later years, Newton's family resented the publicity about Wyatt, especially since they wanted to portray Newton as a successful (and law-abiding) farmer who "remained active in church work, his good deeds never making the newspapers." Quoted in Chaput, *Earp Papers*, 230.

175 *Dr. Shurtleff stayed all day, as did a nurse:* There is some speculation that the "nurse" that Josephine mentions in the Cason manuscript was actually Flood. He may have preferred anonymity, or perhaps Josephine preferred to write a deathbed scene that included only Dr. Shurtleff, an unnamed nurse, and Josephine herself.

176  *Hattie rode with John Flood in the cortege:* Flood to Edward Earp, June 10, 1952. "Mrs. Earp did not attend her husband's funeral (a reason, I shall explain, to you, personally, upon your visit to LA), so I represented her at the funeral, and rode, in the cortege, with her sister Mrs. Emil Lehnhardt, long since deceased." Ragsdale Collection.

176  *there was room for Wyatt and, someday, for Josephine:* The ashes of Josephine and Wyatt are buried in a single plot in the Hills of Eternity cemetery in Colma, California, an area of 1.9 square miles, which was created as an incorporated cemetery area after burials were prohibited in San Francisco in 1900.

## CHAPTER 5: JOSEPHINE'S LAST TRAIL

178  *Fox sent a crew to film the entire event:* Sadly, only a fragment of this remains; the rest of the footage was lost in a plane crash.

179  *inquired when he could expect his copy of Stuart Lake's book:* "When will Wyatt's story be in print and available? I am exceedingly anxious to read it?" U.S. Marshal Mauk to Josephine Marcus Earp, September 17, 1931, Arizona Historical Society, Tucson.

180  *head over to the Welsh home to scrounge for food:* Grace Welsh Spolidoro, interview by Tefertiller, Tefertiller Collection.

184  *"minor thefts of official funds to felonies of scandalous proportions":* Lake to Josephine, April 27, 1929, Lake Collection, Huntington Library.

186  *his broad hint that "the person" was Josephine:* In his letter of October 8, 1939, Leussler states definitively that "Forty-dollar Sadie" Mansfield was not Josephine "Sadie" Marcus. Lake Collection, Huntington Library. There are other arguments to consider: Sadie Mansfield's name appears in the 1882 Tombstone census, which was compiled in July. By then, Wyatt Earp was long gone from Tombstone. It is unlikely that Josephine would have stayed behind to work and even more unlikely that she would return to using her nom de plume as a prostitute, one already linked publicly to Johnny Behan, *after* she and Johnny split up. This (to me) settles the question that Josephine was not the prostitute Sadie Mansfield. Aside from the Mansfield confusion, there has been speculation that Josephine left home as early as 1874 and had been a teenage prostitute. I have found no corroboration for this theory, other than one tantalizing and unsubstantiated reference from Leonard Cason. He told his sister Jeanne Cason Laing that Josephine was no dance-hall girl. He believed that she ran away from home and became a prostitute when she was fifteen years old, then met Behan, and that the Tombstone fight was "all over her." Jeanne Cason Laing asked her brother for evidence about Josephine's teenage years, and when he offered none, she dismissed his comments as speculative. There the matter stands, unless and until additional evidence comes to light. Jeanne Cason Laing interview, Boyer Collection.

186  *"she was known as Sadie Behan":* Lake to Kent, February 13, 1930, Lake Collection, Houghton Library.

187  *Kent brushed off Lake's circumstantial case against Josephine:* Kent to Lake, February 17, 1930, Lake Collection.

187  *"hit her right between the eyes":* Leussler to Lake, March 6, 1929, Lake Collection.

188 *"She is undependable, mentally; unbalanced, psychopathically suspicious":* Lake to Kent, October 15, 1930, Lake Collection.

188 *"We signed the paper as a matter of record for possible publishers":* Lake to Dodge, February 7, 1929, Lake Collection.

189 *"tact, patience, kindliness, and forbearance":* Kent to Lake, October 23, 1930. Lake Collection.

189 *"what your well-meaning but entirely mis-informed friends and advisors tell you":* Lake to Josephine, January 10, 1931, Lake Collection.

190 *Kent accepted Lake's edited manuscript and sent the author a flattering telegram:* Josephine to Lake, February 9, 1931, on Hunsaker and Cosgrove, Attorneys and Counselors at Law, stationery, and telegram from Kent to Lake, March 4, 1931, Lake Collection.

191 Frontier Marshal *was a runaway success:* Lake had just one major sales disappointment: the mighty Book-of-the-Month Club selected *The Epic of America* by James Truslow Adams, the Pulitzer Prize–winning historian who coined the phrase "the American Dream." Interestingly, both *Frontier Marshal* and *Epic of America* are still in print.

192 *"I think it is the best story of the old West":* Clum to Dodge, November 9, 1931, Lake Collection.

192 *He must have been hypnotized by Wyatt*: Raine to Ticknor, January 6, 1932, Lake Collection.

193 *Josephine was losing her circle of allies:* Lake to Kent, January 15, 1933. "That makes 3 [deaths] in a year," Lake noted, in a letter that he signed, "Yours, until Franklin D. Roosevelt deals from a straight deck." Lake Collection.

194 *the book was "dictated to Mr. Lake by my husband":* Josephine to Dr. Sonnichsen, February 13, 1939, Arizona Historical Society, Tucson.

196 *Edna could live comfortably on her inheritance:* By 1958 the oil royalties were more than half of Edna's annual income. She eventually became a respected artist whose works were shown at the Whitney Museum in New York and galleries in Oakland and San Francisco. In 1958 she married artist Lou Siegriest, and she died in 1966 on a sketching trip to Mexico. Until her death, her husband was unaware of her true age and that the family was Jewish. Kirschner interview with Suzanne Westaway.

198 *courage, integrity, and patriotism:* Jeanne Cason Laing compared Lincoln Ellsworth naming his boat *Wyatt Earp* to naming the space shuttle after Wyatt Earp.

200 *Josephine arrived the next day:* This account draws on the biography of Mabel Cason written by her son Walter (this chapter is called "the Josie Earp venture"), and on my interviews with Walter Cason. It also relies on interviews with Walter, Jeanne, and Rae, recorded by Glenn Boyer. Boyer Collection.

203 *"he didn't have the best of principles where women were concerned":* Mabel Cason to Mrs. Merritt Beeson, April 9, 1956, Boyer Collection.

204 *"her big beautiful dark eyes were sparkling":* Boyer, interview with Jeanne Cason Laing, 1974, Boyer Collection.

206 *"We all loved Aunt Allie because she had such wonderful stories and an Irish wit":* Casey Tefertiller, taped interview with Frank and Barbara Waters, Tefertiller Collection.

206 *"a typical little Jewess":* It is worth noting that Halliwell made these repellent comments not in the 1930s but in 1971. Halliwell continued, "Josephine's father was a Jewish silk mer-

chant in the early days in San Francisco and she was just a typical little east side Jewess. . . .
She got money for anything she did and she never let her right hand know what her left
hand was doing. She was a clever little schemer." Recorded interview, September 21, 1971,
Special Collections, University of Arizona Library.

207  *"The tombstones are rattling all over the place":* Naomi Waters to Frank Waters, 1938, Jeff
Wheat Collection.

208  *"Wyatt Earp's widow was a problem in herself":* Lake to Stanley Rinehart Jr., September 19,
1949, Lake Collection.

208  *It was Josephine's first visit to Tombstone:* This account is drawn from Boyer's recorded interviews with the Cason family, my interview with Walter Cason, and the Cason manuscript.

209  *"Whoring!":* This may have been the source of Leonard Cason's belief that Josephine had
been a prostitute.

210  *"It ought to be another* Gone with the Wind*":* Josephine Marcus Earp to Harrison Leussler,
April 24, 1937, Boyer Collection. *Gone with the Wind* was published in 1936.

212  *"You are perfectly right in telling the story of your life":* Harrison Leussler to Josephine, June
11, 1937, Gary Greene Collection. I have searched for but not located Mabel Cason's original drawings.

213  *Leussler immediately sent their draft chapter to Stuart Lake:* Lake told Mabel that he had seen
her manuscript, and that he believed he had exclusive rights to all future Earp biographies.
Walter Cason unpublished biography of Mabel Cason.

213  *The fire was prepared:* Jeanne Cason Laing describes this as a fireplace, Walter Cason as a
backyard incinerator.

214  *Edna relented and made a small settlement upon her aunt:* Josephine to Flood, September 28,
1940. "I had a letter from Edna three weeks ago last Thursday saying they would give me
thirty dollars a month and that I should withdraw the suit and in the same letter Edna wrote
me Emil enclosed a note to me telling me that they would settle the oil lease with Getty and
they would give me $40 a month back pay." Ragsdale Collection.

216  *he finally had the real story of Tombstone:* Lake to Ticknor, February 24, 1945, Lake Collection.

216  *the family of Mrs. Addie Schofield Sinclair:* The Sinclair family lived in Kern County; presumably, Josephine met Addie through the oil business.


CHAPTER 6: PLANET EARP

221  *making movies and publishing books:* For an extensive roundup of books and films related to
the Earps, see Paul Andrew Hutton, "Showdown at the Hollywood Corral: Wyatt Earp
and the Movies," *Montana: the Magazine of Western History,* Vol. 45, No. 3 (Summer , 1995),
2–31.

221  *some 100,000 American soldiers received free paperbacks:* Lake never stopped being his own
best press agent. After the war, he asked Houghton Mifflin to send a first edition to the
White House at the personal request of President Eisenhower, who (Lake says) became
interested in the book while playing golf with some of Lake's friends.

222 *old steamer trunks in his mother's basement labeled "Property of Wyatt Earp":* Kirschner interview with Felton Macartney.

222 *John Gilchriese, a devoted historian and collector of all things western:* Kirschner interviews with Walter Cason and Murdock Gilchriese.

226 *Gilchriese shared these stories with other people:* See the introduction and notes by William Shillingberg to the catalog of the John D. Gilchriese Collection of Tombstone and the West, auctioned through John's Western Gallery. It is possible that Gilchriese's long-awaited biography, or the notes to it, will someday be published, which will provide an opportunity to verify the stories that Gilchriese claimed to have received from Flood.

227 *"That is what Josie was covering from us":* Mabel to Mrs. William Irvine, February 5, 1959, Boyer Collection.

229 *Allie's irresistibly folksy dialect:* See Waters, *Earp Brothers*.

229 *Waters later admitted that he had combined Allie's words:* See the Frank Waters Papers, Center for Southwest Research, which include correspondence between Waters and the Arizona Historical Society.

233 *This turned out to be . . . John Flood's typed manuscript:* There are many versions and copies of Flood's "original" manuscript. Boyer acquired one that may have been given by Josephine to her niece; he then published this version in a limited edition as *Wyatt S. Earp: Wyatt Earp's Autobiography* (Sierra Vista, Ariz.: Loma V. Bissette, 1981), and deposited the original typescript at the Ford County Historical Society. Flood gave other copies to John Gilchriese. Some of Flood's original notes and drawings became part of the Ragsdale Collection; others are in the Boyer Collection at Ford County Historical Society.

233 *perhaps he had also bowdlerized the memoirs of Josephine Earp:* For an in-depth comparison of Boyer's *I Married Wyatt Earp* to the Cason manuscript, see ladyattheokcorral.com. Other discussions of the Boyer controversy include Jeffrey J. Morey, "The Curious Vendetta of Glenn G. Boyer," *Quarterly of the National Association for Outlaw and Lawman History* 18, no. 4 (October–December 1994): 22–28; Gary L. Roberts, *Trailing an American Mythmaker: History and Glenn G. Boyer's Tombstone Vendetta* (Hamilton, Mont.: Western Outlaw-Lawman History Association, 1998); Gary L. Roberts, "The Real Tombstone Travesty: The Earp Controversy from Bechdolt to Boyer," *Western Outlaw-Lawman History Association (WOLA) Journal* 8, no. 3 (Fall 1999); Tony Ortega, "How the West Was Spun," *Phoenix New Times*, December 24, 1998; and "I Varied Wyatt Earp," *Phoenix New Times*, March 4, 1999. For these and others, see "The Boyer Files," http://www.tombstonehistoryarchives.com/?page_id=80

234 *"If it isn't Josie, it ought to be":* Boyer wasn't the only person who thought that the photograph was authentic: Christenne Welsh and Grace Welsh Spolidoro were among those who had insisted that this photograph was Josephine. Jeanne Laing believed that she herself had given the photograph to Glenn Boyer.

235 *"My mother and Aunt were aware of the earlier 'Clum' manuscript":* Jeanne Cason Laing expressed this idea on many other occasions in recorded interviews. In 1974, when asked by Glenn Boyer about Mabel and Vinnolia's initial reaction to the idea of writing a book

about Josephine's life with Wyatt, Jeanne answered that they were very interested and "they knew this had been done before by Mr. Clum." Boyer fueled the controversy with contradictory accounts of the Clum manuscript over the years, claiming sometimes that he could produce it at will, and at other times that it went missing during his move from Hawaii to Arizona. In more recent comments, he describes the "manuscript" as a "generic term" for source materials.

# | Sources

BOOKS

Alexander, Bob. *John H. Behan: Sacrificed Sheriff.* Silver City, N.M.: High-Lonesome Books, 2002.

Bailey, Lynn R., and Don Chaput. *Cochise County Stalwarts: A Who's Who of the Territorial Years.* Tucson, Ariz.: Westernlore Press, 2000.

Bailey, Lynn R. *Tombstone, Arizona: "Too Tough to Die"; The Rise, Fall, and Resurrection of a Silver Camp, 1878 to 1990.* Tucson, Ariz.: Westernlore Press, 2004.

Barra, Allen. *Inventing Wyatt Earp: His Life and Many Legends.* New York: Carroll & Graf, 1998.

Beach, Rex. *The Spoilers.* New York: Harper & Bros, 1906.

Benjamin, I. J. *Three Years in America, 1859–1862.* Philadelphia: Jewish Publication Society of America, 1956.

Berner, Richard C. *Seattle, 1900–1920: From Boom-town Urban Turbulence, to Restoration.* Pullman: Washington State University Press, 1991.

Boyer, Glenn G. *Suppressed Murder of Wyatt Earp.* San Antonio, Tex.: Naylor, 1967.

Breakenridge, William M. *Helldorado, Bringing the Law to the Mesquite.* Boston: Houghton Mifflin, 1928.

Brooks, Alfred H., George B. Richardson, and Arthur J. Collier. *A Reconnaissance of the Cape Nome and Adjacent Gold Fields of Seward Peninsula, Alaska, in 1900.* Washington, D.C.: U.S. Government Printing Office, 1901.

Brown, Clara S., and Lynn R. Bailey. *Tombstone from a Woman's Point of View: The Correspondence of Clara Spalding Brown, July 7, 1880, to November 14, 1882.* Tucson, Ariz.: Westernlore Press, 1998.

Burns, Walter N., and Will James. *Tombstone: An Iliad of the Southwest.* Garden City, N.Y.: Doubleday, Doran, 1929.

Butler, Anne M. *Daughters of Joy, Sisters of Misery: Prostitutes in the American West, 1865–90.* Urbana: University of Illinois Press, 1985.

Chaput, Donald. *The Earp Papers: In a Brother's Image.* Encampment, Wyo.: Affiliated Writers of America, 1994.

Cole, Terrence, and Jim Walsh. *Nome: City of the Golden Beaches.* Anchorage: Alaska Geographic Society, 1984.

Dempsey, David, and Raymond P. Baldwin. *The Triumphs and Trials of Lotta Crabtree.* New York: Morrow, 1968.

Diner, Hasia R. *A Time for Gathering: The Second Migration, 1820–1880.* Baltimore: Johns Hopkins University Press, 1992.

Earp, Josephine S. M., and Glenn G. Boyer. *I Married Wyatt Earp: The Recollections of Josephine Sarah Marcus Earp.* Tucson: University of Arizona Press, 1976.

Earp, Wyatt, Holliday, J. H., & Stephens, J. R. *Wyatt Earp speaks!: My side of the O.K. Corral shoot-out : plus interviews with Doc Holliday.* Cambria Pines by the Sea, California: Fern Canyon Press, 1998.

Fradkin, Philip L. *The Great Earthquake and Firestorms of 1906: How San Francisco Nearly Destroyed Itself.* Berkeley: University of California Press, 2005.

French, L. H. *Nome Nuggets: Some of the Experiences of a Party of Gold Seekers in Northwestern Alaska in 1900.* New York: Montross, Clarke & Emmons, 1901.

Glasscock, Carl B. *Gold in Them Hills: The Story of the West's Last Wild Mining Days.* Indianapolis: Bobbs-Merrill, 1932.

Guinn, Jeff. *The Last Gunfight: The Real Story of the Shootout at the O.K. Corral and How It Changed the American West.* New York: Simon & Schuster, 2011.

Harrison, E. S. *Nome and Seward Peninsula: A Book of Information about Northwestern Alaska.* Seattle: E. S. Harrison, 1905.

Heilbrun, Carolyn G. *Writing a Woman's Life.* New York: Norton, 1988.

Kahn, Ava F. *Jewish Voices of the California Gold Rush: A Documentary History, 1849–1880*. Detroit: Wayne State University Press, 2002.

Kintop, Jeffrey, and Guy L. Rocha. *The Earps' Last Frontier: Wyatt and Virgil Earp in the Nevada Mining Camps, 1902–1905*. Reno, Nev.: Great Basin Press, 1989.

Lake, Stuart N. *Wyatt Earp: Frontier Marshal*. Boston: Houghton Mifflin, 1931.

Levy, Harriet L., with Mallette Dean. *920 O'Farrell Street*. Garden City, N.Y.: Doubleday, 1947.

Lockley, Fred. *History of the First Free Delivery Service of Mail in Alaska at Nome, Alaska, in 1900*. Seattle: Shorey Book Store, 1966.

Marks, Paula M. *Precious Dust: The American Gold Rush Era, 1848–1900*. New York: Morrow, 1994.

———. *And Die in the West: The Story of the O.K. Corral Gunfight*. New York: Morrow, 1989.

McBride, Joseph. *Searching for John Ford: A Life*. New York: St. Martin's Press, 2001.

Palenske, Garner A., and Ben T. Traywick. *Wyatt Earp in San Diego: Life after Tombstone*. Santa Ana, Calif.: Graphic, 2011.

Parsons, George W., and Lynn R. Bailey. *A Tenderfoot in Tombstone: The Private Journal of George Whitwell Parsons; The Turbulent Years, 1880–82*. Tucson, Ariz.: Westernlore Press, 1996.

Pogrebin, Abigail. *Stars of David: Prominent Jews Talk about Being Jewish*. New York: Broadway Books, 2005.

Pourade, Richard F. *The History of San Diego*. San Diego, Calif.: Union-Tribune, 1960.

Reidhead, S. J. *A Church for Helldorado: The 1882 Tombstone Diary of Endicott Peabody and the Building of St. Paul's Episcopal Church*. Roswell, N.M.: Jinglebob Press, 2006.

Rischin, Moses, and John Livingston. *Jews of the American West*. Detroit: Wayne State University Press, 1991.

Roberts, Gary L. *Doc Holliday: The Life and Legend*. Hoboken, N.J: John Wiley & Sons, 2006.

———. *Trailing an American Mythmaker: History and Glenn G. Boyer's "Tombstone Vendetta."* Hamilton, Mont.: Western Outlaw-Lawman History Association, 1998.

Robins, Elizabeth, Victoria J. Moessner, and Joanne E. Gates. *The Alaska-Klondike Diary of Elizabeth Robins, 1900*. Fairbanks: University of Alaska Press, 1999.

Schlissel, Lillian. *Women's Diaries of the Westward Journey*. New York: Schocken, 1982.

Seagraves, Anne. *Soiled Doves: Prostitution in the Early West*. Hayden, Idaho: Wesanne, 1994.

Shillingberg, William B. *Dodge City: The Early Years, 1872–1886*. Norman, Okla.: Arthur H. Clark, 2009.

———. *Tombstone, A.T.: A History of Early Mining, Milling, and Mayhem*. Spokane, Wash.: Arthur H. Clark, 1999.

Stegner, Wallace. *Angle of Repose*. Garden City, N.Y: Doubleday, 1971.

Tanner, Karen H., and Robert K. DeArment. *Doc Holliday: A Family Portrait*. Norman: University of Oklahoma Press, 1998.

Tefertiller, Casey. *Wyatt Earp: The Life behind the Legend*. New York: J. Wiley, 1997.

Traywick, Ben T. *Behind the Red Lights*. Tombstone, Ariz.: Red Marie's, 1993.

Turner, Frederick J., Everett E. Edwards, and Fulmer Mood. *The Early Writings of Frederick Jackson Turner: With a List of All His Works*. Madison: University of Wisconsin Press, 1938.

Turner, Frederick J., and John M. Faragher. *Rereading Frederick Jackson Turner: The Significance of the Frontier in American History, and Other Essays*. New York, N.Y.: H. Holt, 1994.

Walsh, Raoul. *Each Man in His Time: The Life Story of a Director*. New York: Farrar, Straus and Giroux, 1974.

Waters, Frank. *The Earp Brothers of Tombstone: The Story of Mrs. Virgil Earp*. New York: C. N. Potter, 1960.

Wells, E. H., and Randall M. Dodd. *Magnificence and Misery: A Firsthand Account of the 1897 Klondike Gold Rush*. Garden City, N.Y.: Doubleday, 1984.

NEWSPAPER AND JOURNAL ARTICLES

Carlson, T. H. "The Discovery of Gold at Nome, Alaska." *Pacific Historical Review* 15, no. 3 (1946): 259–78.

Chandler, Robert J. "A Smoking Gun? Did Wells Fargo Pay Wyatt Earp to Kill Curly Bill and Frank Stilwell? New Evidence Seems to Indicate Yes." *True West*, July 2001.

———. "Under Cover for Wells Fargo: A Review Essay." *Journal of Arizona History* 41 (Spring 2000).

———. "Wells Fargo and the Earp Brothers: The Cash Books Talk." *California Historical Quarterly*, no. 78 (Summer 2009).

Collier, Anne E. "Harriet 'Hattie' Catchim: A Controversial Earp Family Member." *Western Outlaw-Lawman History Association (WOLA) Journal* 16, no. 2 (Summer 2007).

———. "Stuart N. Lake's Wyatt Earp and the Great Depression." B.A. thesis, University of La Verne, 2011.

Dworkin, Mark. "Wyatt Earp's 1897 Yuma and Cibola Sojourns: Wyatt Earp Returned to Arizona Numerous Times after 1882 Despite Stillwell and Cruz Murder Warrants." *Western Outlaw-Lawman History Association (WOLA) Journal* 14, no. 1 (Summer 2005).

———. "Henry Jaffa and Wyatt Earp: Wyatt Earp's Jewish Connection; A Portrait of Henry Jaffa, Albuquerque's First Mayor." *New Mexico Jewish Historical Society* 19, no. 4 (December 2005).

Earl, Phillip I. "Tex Rickard—The Most Dynamic Fight Promoter in History." *Boxing Insider*, April 15, 2008. http://www.boxinginsider.com/history/tex-rickard-the-most-dynamic-fight-promoter-in-history.

Gatto, Steve. "Wyatt Earp Was a Pimp." *True West*, July 2003.

Hutton, Paul Andrew. "Showdown at the Hollywood Corral: Wyatt Earp and the Movies," *Montana: The Magazine of Western History*, Vol. 45, No. 3 (Summer, 1995), 2-31.

Isenberg, Andrew C. "The Code of the West: Sexuality, Homosociality, and Wyatt Earp." *Western Historical Quarterly* 40, no. 2 (2009): 139.

Hornung, Chuck, and Dr. Gary L. Roberts. "The Split: Did Doc and Wyatt Split Because of a Racial Slur?" *True West*, December 2001.

Kahn, Ava F. and Eisenberg, Ellen. "Western Reality: Jewish Diversity Through the 'German' Period." *American Jewish History*, Volume 92, No. 4 (December, 2004): 455–479.

Jay, Roger. "Wyatt Earp's Lost Year." *Wild West* 16, no. 2 (August 2003): 46.

———. "Reign of the Rough-Scruff: Law and Lucre in Wichita." *Wild West* 18, no. 3 (October 2005): 22.

Mayer, Carl J. "The 1872 Mining Law: Historical Origins of the Discovery Rule." *The University of Chicago Law Review* 53, no. 2 (1986): 624–53.

Mitchell, Carol. "Lady Sadie." *True West*, February/March 2001, 58.

Morey, Jeff. " 'Blaze Away!' Doc Holliday's Role in the West's Most Famous Gunfight." http://home.earthlink.net/~knuthcol/Itemsofinterest4/blazeawaysource.htm.

————. "The Curious Vendetta of Glenn G. Boyer." *Quarterly of the National Association for Outlaw and Lawman History (NOLA)* 18, no. 4 (October–December 1994): 22–28.

Ortega, Tony. "How the West Was Spun." *Phoenix New Times*, December 24–30, 1998. http://www.phoenixnewtimes.com/1998-12-24/news/how-the-west-was-spun/.

————. "I Varied Wyatt Earp." *Phoenix New Times*, March 4, 1999. http://www.phoenixnewtimes.com/1999-03-04/news/i-varied-wyatt-earp/.

Potter, Pamela. "Wyatt Earp in Seattle," *Wild West* 20, no. 3 (October 2007): 46.

Prescott, Cynthia. " 'Why She Didn't Marry Him': Love, Power, and Marital Choices on the Far Western Frontier." *Western Historical Quarterly* 38, no. 1 (Spring 2007).

Raine, William M. "Helldorado: Stories of Arizona's Wild Old Days, When You Couldn't Keep a Bad Man Down." *Liberty*, July 16, 1927.

Roberts, Gary L. "The Real Tombstone Travesty: The Earp Controversy from Bechdolt to Boyer." *Western Outlaw-Lawman History Association (WOLA) Journal* 8, no. 3 (Fall 1999).

————. *Trailing an American Mythmaker: History and Glenn G. Boyer's "Tombstone Vendetta."* Hamilton, Mont.: Western Outlaw-Lawman History Association, 1998.

————. "Allie's Story: Mrs. Virgil Earp and the *Tombstone Travesty*." http://home.earthlink.net/~knuthcol/Travesty/AlliesStory1source.htm.

Shillingberg, William. "Wyatt Earp and the 'Buntline Special' Myth." *Kansas Historical Quarterlies* 42, no. 2 (Summer 1976): 113–54.

————. Introduction and notes to the catalog for the John Gilchriese Collection, John's Western Gallery, 2004.

St. Johns, Adela Rogers. "I Knew Wyatt Earp." *American Weekly*, May 22, 1960 reprinted in John Richard Stephens, ed., *Wyatt Earp Speaks!*, Cambria, CA: Fern Canyon Press, 1998.

Tefertiller, Casey, and Jeff Morey. "O.K. Corral: A Gunfight Shrouded in Mystery." *Wild West* 14, no. 3 (October 2001): 48.

————. "What Was Not in *Tombstone Travesty*." http://home.earthlink.net/~knuthcol/Travesty/notintravestysource.htm.

UNPUBLISHED MANUSCRIPTS AND COLLECTIONS

Various documents, Glenn Boyer Collection, Special Collections, University of Arizona. Made available by permission of Glenn Boyer.

Various documents and photographs, Boyer Collection. Made available by permission of Glenn Boyer.

Cason manuscript (original), Special Collections, Ford County Historical Society. Made available by permission of Glenn Boyer.

John Clum Collection, Special Collections, University of Arizona.

Various documents, Scott Dyke Collection. Made available by permission of Scott Dyke.

Louisa Houston Earp letters, Special Collections, Ford County Historical Society. Made available by permission of Glenn Boyer.

John Flood interview notes, letters of Josephine Marcus Earp, and various documents and photographs, Ragsdale Collection. Made available by Mark Ragsdale.

Various documents, Gary Greene Collection. Made available by Gary Greene.

Various documents, Stuart N. Lake Collection, Huntington Library, San Marino, California.

"The Private Journal of George Whitwell Parsons, January 1, 1880–June 28, 1882." Edited by Carl Chafin. Made available by Christine Rhodes.

Avaloo Boyd, "Alaska, I Love You," unpublished memoir. Carrie M. McLain Memorial Museum, Nome, Alaska.

Various documents, Ellsworth Collection, Arizona Historical Society, Tucson, Arizona.

Edna Lehnhardt Cowing Stoddard Siegriest diaries. Made available by Suzanne Westaway.

Various documents and letters, Houghton Mifflin Collection, Houghton Library, Harvard University.

Celia Earp Inquest, Arizona Department of Library, Archives Division, Pinal County Inquests, /filmfile 88.6.1.

Mrs. John F. Mercer, Nome trip scrapbook, Archives and Special Collections, Consortium Library, University of Alaska, Anchorage.

Herbert Heller Papers, Lynn Smith Correspondence and Letters, University of Alaska, Fairbanks.

Alaska Commercial Company Records, Green Library, Stanford University, California.

Various documents, Western Jewish History Center, Judah L. Magnes Museum, University of California, Berkeley.

Burial records, Hills of Eternity cemetery, Colma, California.

## RECORDINGS

Greenberg, Alice "Peggy," and Alvin Greenberg. Taped interview. Collection of Roger S. Peterson.

Greenberg, Alice "Peggy," Alvin Greenberg, and Jacqueline Wolf. Taped interviews. Collection of Glenn Boyer.

Laing, Jeanne Cason. Taped interview. Collection of Roger S. Peterson.

Laing, Jeanne Cason, Walter Cason, and Rae Cason Lindsay. Taped interviews. Collection of Glenn Boyer.

Halliwell, Hildreth. Taped interview. Collection of Glenn Boyer.

———. Taped interview. Special Collections, University of Arizona.

Hart, William S., Jr. Taped interview. Collection of Roger S. Peterson.

Macartney, Marjorie. Taped interview. Collection of Glenn Boyer.

Ojala, Arvo. Taped interview. Collection of Casey Tefertiller.

Siegriest, Louis. Taped interview. Collection of Glenn Boyer.

Waters, Frank, and Barbara Waters. Taped interview. Collection of Casey Tefertiller.

Welsh, Christenne, Grace Welsh Spolidoro, and Elena Welsh Armstrong. Taped interviews. Collection of Casey Tefertiller.


## WEB SITES

B.J.'s Tombstone Discussion Forum. http://disc.yourwebapps.com/Indices/39627.html.

Legal and Court History of Cochise County. http://azmemory.azlibrary.gov/cdm/search/collection/ccolch.

Tombstone History Archives. http://www.tombstonehistoryarchives.com/

# | Photo Credits

# | Index

Ackerman, Harold, 201, 205, 208, 209, 211, 214, 222, 223, 226–27

Ackerman, Vinnolia Earp, 8, 10, 11, 137, 201–3, 205, 208–15, 217, 218, 222, 223, 225, 233

*Adventure Magazine*, 164

Alaska, 92–134. *See also* Nome, Alaska; Rampart, Alaska; St. Michael, Alaska

Bering Sea, 105, 108, 111, 115, 120

boomtowns of, 93, 107, 122, 132–33 (*see also* Nome, Alaska)

Chilkoot Pass, 94, 95, 128

Earps in Nome, 107–15

Earps in Rampart, 96–104

Earps in St. Michael, Alaska, 104–7

Earps' trip up the Yukon river, 95–96

Jewish financing of Alaska mining supplies and ventures, 96 (*see also* Alaska Commercial Company)

Klondike gold fields, 89, 92, 93, 94, 103, 107

miners and hardships of, 94

prices of goods in, 105

psychological hardships in, 99

suicide in, 99–100, 254n 100

travel to, 93

weather of, 102–3

winters in, 99, 114–15, 117–18

Alaska Commercial Company, 96, 125–26

in Nome, Alaska, 112, 123–24

in St. Michael, Alaska, 105

Wyatt Earp and, 104, 105–6

*Alliance* (ship), 119–20, 255n 119

American West. *See also specific cities and states*

boomtowns of, 9, 29, 42, 75, 93, 135, 138, 256n 136

common-law marriages in, 37–38, 166–67, 249n 38

ending of the frontier, 3, 87

in films, 151–52, 153, 221, 234

gamblers and, 42, 149

American West *(cont.)*
  *Harper's* short story series and, 87–88
  Jewish entrepreneurs in, 42, 70
  lawmen of, 42
  "manifest destiny," 34
  prostitution in, 7, 36, 42, 47–49, 82,
    84–85, 127, 137, 201, 211, 250n 49,
    258n 186, 260n 209
  Red Mountain Pass, 72–73
  Stegner on, 8
  television shows about, 222
  travel in, 13
  travel to California, 1860 to 1870, 17
  wagontrains to, 34–35, 249n 34
  women's lives in, 7, 41
  women unacknowledged in histories,
    6, 50, 164
  women with occupations, 47
  writings about, 87–88, 175, 183, 199–
    200, 221 (*see also* Boyer, Glenn; *Frontier
    Marshal; Helldorado; Sunset Trail, The*)
  Wyatt Earp as legend, 7, 11, 73, 75, 229
Apaches, 52, 130
*Arctic Weekly Sun*, 128
Arizona
  boomtowns of, 29
  Civil War and divisiveness in, 50
  Cochise County, 7, 43, 50
  at Columbian Exhibition, 87
  Indian hostilities in, 52
  lawlessness in, 61
  Prohibition in, 144
  Tombstone as largest city, 28
Arizona Historical Society, 229
Arizona Pioneers Historical Society, 168
Arthur, Chester, 61

Bakersfield, California, 146
Baldwin, Elias Johnson (Lucky), 20,
  80–81, 85, 86, 87, 90, 130–31, 132,
  144, 256n 136

  death of, 142
  Josephine Earp's wedding ceremony
    story on yacht of, 85, 253n 85
Barra, Allen, 234
*Bat Masterson* (TV series), 222
Beach, Rex, 102, 107, 108, 113, 118, 125,
  134, 170, 254n 97
Bechdolt, Frederick, 6, 164, 167, 168
Behan, Albert, 31, 48, 49, 94–95, 165,
  169, 208, 209, 257n 165
Behan, Johnny
  Benson stagecoach attackers and, 51
  Bisbee stagecoach attackers and, 52
  Burns' description of, 164
  business ventures in Tombstone, 31,
    32
  career after Tombstone, 65–66
  character and personality, 44, 65, 79
  child and marriage of, 25, 26, 28, 30
  Clantons and McLaurys and, 45, 58
  as corrupt, 52, 53, 57
  cowboys, rustlers, and, 44–45, 52, 53,
    57
  death of, 66, 148
  as Democrat, 32, 44, 51
  as Deputy Sheriff, 25
  Earp brothers as enemies of, 44
  Earp-cowboy war and, 59
  end of law career, 65
  Gunfight at the O.K. Corral and, 7, 56,
    148, 154, 179
  Ike Clanton and, 58
  indictment of Earp for Stilwell murder
    and, 70–71
  Josephine Marcus and, 210, 212, 250n
    46
  lead-up to Gunfight at the O.K. Corral,
    54
  meets Josephine Marcus, 25
  relationship with Josephine Marcus
    ends, 46–47, 48

relationship with Josephine Marcus in
  Tombstone, 7, 30, 31–32, 41–42, 45,
  185
rivalry with Earp, 7, 43–44, 49, 51, 64,
  159, 252n 74
as Sheriff, Cochise County, 7, 43–44,
  59, 65
Spence as associate of, 52
Stilwell as associate of, 52
syphilis and, 46–47, 250n 46
testimony at trial following Gunfight,
  58, 59
as womanizer, 30–31, 46, 186, 250n 46
wooing of Josephine Marcus, 26–28
Behan, Victoria Zaff, 30–31
Belasco, David, 20, 21, 248n 21
*Belle of the Yukon* (film), 214
Benjamin, Israel Joseph, 18–19
Benjamin of Tudela, 18
Benson, Arizona, 13, 28
  stagecoach robbery and murder, 51–52
Bisbee, Arizona, 43, 177
  stagecoach robbery, 52
Black Thursday and stock market crash
  of 1929, 179
Blaylock, Celia "Mattie" (see Earp, Mat-
  tie Blaylock)
"Blond Beauties" (Thompson), 23
Borland, Addie, 47, 63, 190
Boyer, Glenn, 11, 230–37, 262n 233, 262n
  234
  Clum manuscript and, 235–36
Brainerd, Erasmus, 100, 116, 254n 100
Breakenridge, William "Billy," 6, 50, 65,
  165–69, 178, 186, 192, 200
Brown, Clara Spalding, 39–40
  on "Cowboy and Earp Feud," 60
  on cowboys, 43
  departure from Tombstone, 67
  fire in Tombstone and, 45
  Gunfight at the O.K. Corral described, 55

report on cowboys' funeral after Gunfight, 56–57, 59
reports to *San Diego Union*, 39–40,
  45–46, 47
on saloons and brothels, 48
on Stilwell killing, 64–65
Burnett, William R., 184
Burns, Walter Noble, 6, 44, 50, 161–62,
  167, 168, 178
  description of Wyatt Earp, 33

Cashman, Nellie, 47, 200
Cason, Ernest, 201, 205, 214, 227, 231,
  236
Cason, Leonard, 204, 205, 231
Cason, Mabel Earp, 8, 10, 11, 127, 137,
  201–3, 208–15, 218, 222–27, 231, 233,
  236, 248n 19, 260n 213
Cason, Rae, 204, 205, 215
Cason, Walter, 8–9, 11, 203, 222
Chaplin, Charlie, 152
*City of Seattle* (ship), 93
Civil War
  common-law wives and pensions, 38,
  139
  divisiveness in Tombstone and, 50
  Earp brothers in, 34, 50
Claiborne, Billy, 55
Clanton, Billy, 54, 56
Clanton, Ike, 59
  charges and trial following Gunfight,
  57–59
  Earp bribes into turning over Benson
  killers, 52, 58
  Earp-cowboy war and, 61
  Earp saves him during Gunfight at the
  O.K. Corral, 56, 59
  Gunfight at the O.K. Corral, 55, 56
  lawsuit against Earp, 59–60
  lead-up to Gunfight at the O.K. Corral,
  53–54

Clanton, Phin, 251n 63
*Cleveland* (ship), 115
Cleveland, Grover, 87
Cline, Edward F., 215
Clum, John, 29, 45, 59, 65, 148, 156,
    165, 168–69, 173, 175, 178, 187, 192,
    199–200, 202, 208, 247n 13
 in Alaska, 128, 255n 128
 death of, 193
 description of Wyatt Earp, 33
 manuscript, 231, 235–36, 262n 234
Clum, Mary Ware, 247n 13
Coburn, Walter, 170, 215–16
Cody, William "Buffalo Bill," 87, 201
Coeur d'Alene, Idaho
 Earp brothers in, 75
 White Elephant saloon, 75
 Wyatt Earp as deputy sheriff, 75
Cohn, Alice, 218, 219
Colton, California, 74
Columbian Exhibition (1893), 86–88
 Buffalo Bill Cody at, 87
 Earp's horse races at, 87
cowboys
 appearance of, 10
 Behan and, 7, 52, 53
 "buffaloing" of, 36
 cattle thievery in Tombstone and,
    44–45
 communities' reaction to, 36
 defined as villains, rustlers, outlaws, 43
 Earp-cowboy war, 59–64
 funeral in Tombstone, 56–57
 Gunfight at the O.K. Corral and,
    55–56, 179
 in Tombstone, 10, 43, 169
 Wister's *The Virginian*, 88
Cowing, Estes Joseph, 145
Cowing, Marjorie (Lehnhardt), also
    Marjorie Macartney, 145, 148
Crabtree, Annie Leopold, 166, 208

Crabtree, Jack, 166
Crabtree, Lotta, 166
Crummy, George W., 71, 252n 71

*Daily Alta California* newspaper, 17
Dawson, Alaska, 95, 102, 108, 125–26
DeMille, Cecil B., 151
Dempsey, Jack, 113
Denver, Colorado
 Doc Holliday in, 76
 Tabor or Windsor Hotel, 72, 76, 252n 72
 Wyatt and Josephine Earp in, 72, 76, 87
*Denver Republican*, 87
Dexter, John, 113
Dodge, Fred, 178, 187, 192, 200
Dodge City, Kansas, 36, 37, 152, 182, 208
 Doc Holliday and Wyatt Earp in, 37
 Ford County Historical Society, 10
 Historical Society, 236–37
 Long Branch saloon, 73
 museum, 227
 War Peace Commission, 73
 Wyatt Earp returns, 1883, 73–74
*Dodge City Globe*, 192
Doheny, Edward, 171, 176, 181, 214, 257n
    171
Dubin, Rabbi Maxwell, 218
Dunbar, Emma, 46
Dunbar, John O., 31, 32, 41–42, 46
Dwan, Alan, 152, 153

Earp, Alvira "Allie" Packingham Sul-
    livan, 38, 56, 61, 62, 77, 138–40,
    205–6, 209
 on Josephine Earp, 50
 life of, 41
 memorabilia provided by, 183
 opinion of Wyatt, 41
 sewing machine and money earned, 40
 Waters's book and, 206, 207–8, 228–29
Earp, Aurilla Sutherland, 35, 211

Earp, Bessie, 37, 63
  daughter of, 42, 250n 42
  as former madam, 36, 42
Earp, Celia "Mattie" Blaylock
  as common-law wife of Wyatt Earp,
    37, 38, 41, 56, 60, 63–64, 71, 164, 185,
    186, 187, 211, 249n 37
  death of, 84–85, 206
  truth emerges about, 227–28, 236
Earp, Hattie (Catchim), 42, 250n
Earp, James, 34
  brothels in Peoria, 36
  common-law wife Bessie, 36, 37, 42, 63
Earp, Josephine Sarah Marcus, 27–28,
    197
  as actress and runaway, 1879, 13, 14,
    22–26, 145, 186, 210, 212
  appearance, 4, 5, 13, 14, 75, 136–37,
    140, 141, 200, 214–15
  Bat Masterson and, 72
  birth of, 3, 5, 15, 247n 5
  Breakenridge's book about Wyatt Earp
    and, 169
  Burns' book about Wyatt Earp and,
    161–63
  canary, "Dickie," 73
  character and personality, 4, 6, 19–20,
    22, 91–92, 102, 104, 143, 147–48, 176,
    180, 181, 187–88, 190, 203, 205, 211,
    223, 224, 237–38
  childhood in San Francisco, 18–21
  childhood journey from New York to
    San Francisco, 18
  at Columbian Exhibition, 86–87
  death of, 218
  death of sister Hattie, 195, 196, 197, 215
  Doc Holliday and, 70, 76
  fabrications by, 3, 5, 8, 11, 18, 26–27,
    85, 210–11, 236
  family and, 3–4, 14–20, 147–48, 197
  final possessions, 217

Flood's biography of Wyatt Earp and,
    154–61
  gambling and, 4, 8, 81, 83, 127, 141, 149,
    180, 255n 120, 257n 171
  grief over Wyatt's death, 179, 181
  historical sources for, 2, 7, 10–12
  income and finances, 31, 32, 46, 124, 132,
    135, 146, 170, 171, 180–81, 192–93,
    199, 202, 214, 217, 218–19, 257n 171
  Johnny Behan wooing of, 25–28
  Judaism of, 1–2, 4, 21, 42, 176, 204, 218
  Lake biography of Wyatt Earp and,
    171, 172–73, 182–94, 210
  last years of, 179–219
  lawsuit against niece over oil royalties,
    196–97, 205, 214, 259n 196, 260n 214
  love of adventure, 5, 14, 21, 24, 91–92, 112
  love of children, 94
  Lucky Baldwin and, 20, 80–81, 129–30
  memoirs of, 8, 10–11, 63, 200–203,
    208–16, 222–27, 248n 19, 260n 213
  memorabilia of Wyatt Earp's and, 198,
    216
  mining claims, 139, 180–81, 205, 207, 216
  obituary, 218
  as "Sadie" vs. "Josie," 5–6
  in San Francisco, 1880, 27
  inaccurate accounts about, 225–26, 229
  secrets of, 7–8, 26, 84–85, 142, 164,
    185, 199, 206, 213, 227
  self-invention, 9, 21
  shaping of public face and legend of
    Wyatt Earp, 7, 35, 150–51, 154–61,
    171, 179, 183–94, 210–13, 215–16, 238
  siblings, 15, 69, 142
  singing and dancing lessons, 23, 210
  social discrimination and, 21, 41–42,
    106, 125
  as stagestruck, 20, 23
  St. Vitus's dance illness, 27
  will of, 197, 216–19

Earp, Josephine Sarah Marcus *(cont.)*
Tombstone and, 7
  affair with Wyatt Earp, 49–50, 63, 186
  arrival, 1880, 13–14, 28–30, 247n 13
  Behan and, 7, 28, 30, 31–32, 41–42,
    43–44, 45, 164, 185, 210, 212
  Behan relationship ends, 46–47, 48,
    250n 46
  Behan's son Albert and, 30, 31, 48,
    94–95, 165, 210
  departure from, 7, 63, 251n 63
  Gunfight at the O.K. Corral and, 3,
    6–7, 10, 12, 56, 58–59
  house in, 32, 250n 46
  meets Wyatt Earp, 49
  prostitution suspicions, 48–49, 186,
    209, 258n 186, 260n 209
  return to, 1937, 208–9
  shadow of events, on her life, 7–8, 74,
    127, 142, 148
  social status, 41–42
Wyatt Earp years, 89
  in Alaska, 93–134
  begins using name "Mrs. Earp," 1883,
    74, 252n 74
  camp at Vidal, California, 140, 165,
    173, 194
  crossing Red Mountain Pass, 72–73
  defense of Wyatt, 128, 143, 148, 150, 154
  devotion to Wyatt, 202
  dislike of his "sporting life," 113, 137
  domestic disputes, 82–83, 203
  first home, Rampart, Alaska, 97–98
  gifts from Wyatt, 79
  in Goldfield, Nevada, 138–40
  horseracing and, 79–82, 84, 85–86
  jealousy and other women, 82, 202–3
  length of marriage, 7, 9
  in Los Angeles, Wyatt's final years and
    death, 151–73
  marital tensions, Nome, 126–27

marriage as common-law, 4, 71, 92
nomadic life, 74–76, 135–44
in Nome, Alaska, 107–15
parents' and, 71
pregnancy and miscarriage, 94, 95, 101,
  102, 253n 94, 254n 94
in Rampart, Alaska, 96–104, 254n 97
in San Diego, 76–84, 252n 82
in San Francisco, 1882, waiting for
  Earp, 68, 69–71
in San Francisco, living with Earp,
  85–88, 90–91
in Seattle, Washington, 116–17
separation, summer of 1883, 73–74
in St. Michael, Alaska, 104–7
in Texas, 75
in Tonopah, Nevada, 135–38
in Utah, Colorado, and Idaho, 71–73
wagons and supplies, life on the road,
  136–37
wedding ceremony story by, 85, 253n
  85, 253n 94
Wyatt's infidelity, 127, 202–3
in Yuma, Arizona, 89–90, 92
Earp, Louisa Houston, 37, 38, 56, 60,
  249n 38
  Apache threat described by, 52
Earp, Morgan, 37, 38, 156
  common-law wife Louisa, 37, 38, 249n
    38
  Gunfight at the O.K. Corral, 55, 56
  killing of, 61–62, 208–9, 251n 61, 251n
    63
  lead-up to Gunfight at the O.K. Corral,
    54
  as shotgun messenger for Wells Fargo,
    28
Earp, Newton, 34
  death of, 174, 257n 174
Earp, Nicholas, 34, 37, 38
  death of, 142

Earp, Sally. *See* Haspel, Sally
Earp, Virgil, 32, 33, 34, 37, 41, 53, 128
  Benson stagecoach attackers sought by,
    51–52
  common-law wife Alvira "Allie," 38,
    41, 61, 62, 206
  death of, 139, 256n 139
  as deputy U.S. marshal, 38, 51
  in Goldfield, Nevada, 138–39
  Gunfight at the O.K. Corral, 55, 56
  Happy Days mines and, 139, 181, 207
  lead-up to Gunfight at the O.K. Corral,
    53–54
  in San Diego, 77
  in San Francisco, 71
  shooting of, 61, 208
  wife, Jane Sysdam, 256n 139
Earp, Virginia Cooksey, 34, 37, 38
Earp, Warren, 127–28, 255n 119
Earp, William Harrison, 201
Earp, Wyatt, 126–27
  accident in San Francisco, 93, 94
  Alaska Commercial Company and, 104,
    105–6
  appearance, 5, 10, 32–33, 71, 87,
    140–41, 152, 156
  as archetype, 7
  arrests in Peoria, 36
  Bat Masterson and, 7, 33, 35, 73, 78–79,
    148–49, 154
  biography by Flood, 154–61, 171–72,
    233, 261n 232
  biography by Lake, 170–73, 183–94
    (*see also Frontier Marshal*)
  birth of, 34
  Breakenridge book about, 165–69 (*see
    also Helldorado*)
  brothers of, 2, 24, 32–34, 41, 174 (*see
    also* Earp, Morgan; Earp, Virgil)
  brothers' wives and, 33
  bunco scheme arrest, 142–44

  Burns' book about, 161–63, 168
  carrying a weapon and, 90, 91
  character and personality, 7, 9, 32, 33,
    41, 44, 72, 78–79, 147, 148, 156, 170,
    172, 228, 230, 256n 148
  Cochise County sheriff position sought
    by, 51
  common-law wife Mattie Blaylock,
    37, 38, 41, 56, 60, 63–64, 68, 71,
    74–75, 84–85, 185, 186, 187, 211, 212,
    227–28, 249n 37
  common-law wife Sally Haspel, 36,
    249n 36
  dark years following Aurilla's death,
    35–36
  death and final year of, 170–75
  death of brother, Morgan, 61–62, 251n
    61, 251n 63
  death of brother, Newton, 174, 257n 174
  death of brother, Virgil, 139, 256n 139
  death of brother, Warren, 127–28, 255n
    119
  definition of justice, 79
  Doc Holliday and, 37, 70, 76, 156, 249n
    37
  fame of, 7, 8, 113, 117, 124, 135, 143,
    197–99, 221
  family and ethnic heritage, 34–35
  fight-fixing scandal, 90–91, 117, 143,
    253n 91
  Flood as legal advisor, 143–44, 155,
    180–82 (*see also* Flood, John)
  friends of, 7, 103, 140, 153, 171, 173,
    175–76, 178, 182, 192, 193, 208 (*see
    also specific people*)
  funeral, 175–76, 258n 176
  gambling and, 32, 35, 37, 39, 60, 76, 77,
    80, 82, 97, 100, 113, 117, 137, 142–43,
    149, 180, 254n 119
  George Hearst and, 43, 75, 151
  grave of, 1, 4, 176, 218, 221, 258n 176

Earp, Wyatt *(cont.)*
Hollywood and, 152–56, 160, 176
horseracing and, 79–82, 84, 85–86, 153
income and finances, 60, 70, 81, 113,
117, 124, 132, 135, 146, 147, 171, 195,
196, 256n 146
in Lamar, Missouri, 35
as legend, 7, 11, 73, 75, 229
as "Lion of Tombstone," 33, 163
Lotta Crabtree estate case and, 166–67
loyalty of, 37, 41, 76, 161
marriage to Aurilla Sutherland, 35, 211
memorabilia, 198, 216
men killed by, 70
police work, 36, 75, 137
portrayed as teetotaler, 158
prizefighting and, 124, 138
prostitution and, 36, 82, 127, 137, 211
as Republican, 44, 50
reputation of, 64, 91, 128, 199
as saloonkeeper, 59, 75, 113, 117, 122,
124, 126, 130, 132, 133, 135–38, 144,
171, 255n 133
inaccurate accounts about, 148, 149, 200
*(see also Helldorado; Pioneer Days in
Arizona)*
ship named for, 197–99
television series about, 222
in Texas, 75
Tom Mix and, 153, 173–74
Wells Fargo Company and, 38–39, 52,
70, 75, 149
Josephine "Sadie" Marcus Earp years,
135
in Alaska, 93–134
builds tree house for Josephine, 140
camp at Vidal, California, 140, 165, 173
domestic disputes, 82–83, 203
final years in Los Angeles, 151–73
gifts to Josephine, 79
in Goldfield, Nevada, 138–40

infidelity and, 127, 202–3
Josephine begins using name "Mrs.
Earp," 1883, 74
Josephine's gambling and, 81–82
length of marriage, 7, 9
in Los Angeles, final years and death,
151–73
marital tensions, Nome, 126–27
marriage as common-law, 4, 71, 92
nomadic life, 74–76
in Nome, Alaska, 107–15
in Rampart, Alaska, 96–104, 254n 97
in San Diego, 76–84, 252n 82
in San Francisco, 85–88, 90–91
in Seattle, Washington, 116–17
separation, summer of 1883, 73–74
in St. Michael, Alaska, 104–7
in Texas, 75
in Tonopah, Nevada, 135–38
treatment of Josephine, 141, 202, 203
in Utah, Colorado, and Idaho, 71–73
in Yuma, Arizona, 89–90, 92
Tombstone years:
affair with Josephine Marcus, 49–50,
63, 186
arrives in Tombstone, 1879, 24, 32
Behan rivalry with, 43–44, 49, 51, 252n
74
Benson stagecoach attackers sought by,
51–52
brother Morgan killed, 61–62
brother Virgil shot, 61
business ventures in Tombstone, 38–39,
59, 65
charges and trial following Gunfight
(Spicer hearing), 57–59
Cochise County sheriff position sought
by, 43–44
departure from, 66
as deputy U.S. marshal, 61
Earp-cowboy war, 59–64

Gunfight at the O.K. Corral, 2–3, 12, 55–56, 66, 127–28, 148
  Ike Clanton and, 52, 56, 58
  killing of Frank Stilwell, 63, 70–71, 154
  lead-up to Gunfight at the O.K. Corral, 51–54
  meets Josephine Marcus, 49
  as Pima County Deputy Sheriff, 51
  as shotgun messenger for Wells Fargo, 38
  Vendetta Ride, 64–66, 70, 148, 154, 168
*Earp Brothers of Tombstone, The* (Waters), 228–29
Edwards, Adelia, 235
Eisenhower, Dwight D., 261n 220
Elder, "Big Nose" Kate, 37, 52, 57, 74, 190, 199, 228, 231
Ellsworth, Lincoln, 197–99
Ephron, Nora, 4
*Epitaph* (newspaper). *See Tombstone Epitaph*

Fairbanks, Douglas, 153
Fitch, Tom, 57–58, 77
Fitzsimmons, Bob, 90–91, 124, 253n 91
Flood, John, 256n 143
  biography of Wyatt Earp, 154–61, 169, 171–72, 233, 261n 232
  as executor and beneficiary, 216, 217–19
  Gilchriese's book and, 224–26
  loyalty of, 180–82, 214, 215
  Wyatt and Josephine Earp and, 11, 143, 147, 148, 149–50, 160, 162–65, 175, 176, 183, 189, 195, 215, 219, 224–25, 258n 175
Ford, John, 152
Frank, A. L., 218
Fremont, John C., 44
*Frontier Marshal* (film), 193–94
*Frontier Marshal* (Lake), 191–94, 197, 199, 213, 215, 228, 230, 231, 259n 191

giveaway to WWII soldiers, 221, 261n 220
new edition for teenage readers, 222
royalties, 182, 192, 193, 195, 197, 199, 207, 216, 221–22

Galveston, Texas, 74, 252n 74
Gannett, Leslie, 192
Geronimo, 130
Gilchriese, John, 222–27, 231, 236, 261n 225
Glenwood, Colorado, 76
Globe, Arizona, 74–75
Goldfield, Nevada, 9, 138–40
  death of Virgil Earp in, 139
  Earp brothers in, 138–40
  New Northern Saloon, 138
  Wyatt and Josephine Earp in, 138–40
  Wyatt and Josephine Earp's claims, Happy Day Mines, 139, 181, 207
*Goldfield News*, 139
Goodfellow, George, 48, 62
Gosper, John, 52–53, 59
Grauman, Sidney, 93, 125, 152, 204, 218
Greenberg, Alice, 233
Groton school, 40
Gump, Solomon, 16

Halliwell, Hildreth, 139, 206–7, 208, 256n 146, anti-Semitism of 260n 206
Hammond, John Hays, 137, 148
Harper, Thomas, 176
*Harper's Monthly Magazine*, 87–88
Hart, William S., 7, 153–54, 155, 160–61, 162, 164, 170, 173, 174, 175, 182, 185, 190, 196, 202, 209, 216, 217, 218
Haspel, Sally, 36, 211, 249n 36
Hatch, Bob, 62, 208
Hatton, Charles, 33
Hearst, George, 7, 43, 75, 151, 215
Hearst, William Randolph, 150–51, 215

*Helldorado* (Breakenridge), 167–69, 173, 200

*Hera* (ship), 115–16

Hirsch, Dora, 23–24

*HMS Pinafore* (Gilbert and Sullivan), 22, 23, 204

Markham "Pinafore on Wheels" Troupe, 24–26, 145, 186, 210

Holliday, John Henry "Doc," 2, 7, 37, 156, 161, 175

Bat Masterson and, 37, 71

"Big Nose" Kate and, 37, 52, 57, 190

Boyer book about, 230, 233, 262n 233

Burns' description of, 163–64

charges and trial following Gunfight, 57–59

Gunfight at the O.K. Corral, 55, 56

illness and death of, 76

implicated in Benson killing, 52

"Jew boy" comment, 71, 252n 71

lead-up to Gunfight at the O.K. Corral, 54

saves Wyatt Earp's life, 37

Wyatt Earp and, 37, 70, 76, 249n 37

Hollywood and film industry, 151–54

Josephine Earp as "technical consultant," 204, 214

westerns, 151–52, 153, 204, 219, 221

Wyatt Earp and, 152–56, 160, 176, 234

Hooker, Forrestine, 155, 231

Hooker, Henry Clay, 155

Hoover, Herbert, 113

Horton, Alonzo, 77, 83

Hotel del Coronado, San Diego, 78

Houghton Mifflin, 167, 168, 169, 178, 182, 191, 209, 212, 216, 231

Josephine Earp at, 189

Hoxsie, Charlie, 113, 114–15, 118, 133, 134

Hunsaker, William J., 77, 83, 148, 172, 175, 179, 193, 252n 77

Hunt, George, 179

Huntington Library, 11

Idaho

boomtown of Coeur d'Alene, 75

Kootenai County, Earp as sheriff, 75

silver fields, 72, 75

*Illustrated Life of Doc Holliday, The* (Boyer), 230, 233

*I Married Wyatt Earp* (unpublished ms.), 10–11

*I Married Wyatt Earp: The Recollections of Josephine Sarah Marcus Earp* (ed. Boyer), 11, 232–34, 262n 233, 262n 234

*Inventing Wyatt Earp* (Barra), 234

Israel, Sol, 42, 49, 65, 156

*It All Happened in Tombstone* (Clum), 199–200

Jackson, Joseph Henry, 192

Jaffa, Henry, 70

Jones, Harry, 27, 30, 42, 48, 49

Jones, Ida "Kitty," 27–28, 30, 42, 48, 208

Judaism

anti-Semitism in America, 6, 206, 260n 206

assimilation and, 14

Eastern European Jews, 14–15, 248n 14

Israel Benjamin as commentator on Jewish life, 18–19

Jewish community of New York City, 15, 248n 14, 248n 16

Jewish community of Nome, Alaska, 125–26

Jewish community of San Francisco, 4, 16–17, 20–21, 248n 16

Jewish community of Tombstone, 42

Jewish financing of Alaska mining supplies and ventures, 96, 125–26

of Josephine Earp, 1–2, 4, 21, 147, 204, 218
in 19th century Prussia, 14–15
prejudices of German Jews against
  Eastern European Jews, 3, 15, 19,
  20–21, 248n 19, 248n 20
Wyatt Earp and, 147

Kelley, Florence Finch, 191–92
Kent, Ira Rich, 168, 187, 189, 190, 193

Labatt, Abraham, 17
Laing, Jeanne Cason, 202–4, 205, 225,
  231–32, 234, 236–37
Lake, Stuart, 228, 230, 257n 170
  Josephine Earp's memoirs and, 213,
  216, 219, 223, 224
  omission of women or wives from *Fron-
  tier Marshal*, 190
  Wyatt and Josephine Earp and *Frontier
  Marshal*, 11, 49, 50, 170–73, 176, 179,
  182, 183–94, 195, 198, 207, 208, 210,
  211, 214, 219, 221–22, 224
Langtry, Lily, 78
*Law and Order* (film), 193
Lazard Frères, 16
Lehnhardt, Edna (niece), 145, 146, 180,
  196–97, 209, 214, 217, 259n 196, 260n
  214
Lehnhardt, Emil (brother-in-law),
  69–70, 71, 74, 91, 142
  suicide of, 145, 209
Lehnhardt, Emil, Jr.(nephew), 137, 196
Lehnhardt, Henrietta "Hattie" (sister),
  15, 20, 69–70, 74, 76, 85, 91, 137, 142
  as businesswoman, 146
  closeness with Josephine, 146–47,
  179–80, 195, 215
  death of, 195, 196, 197
  Earp's funeral and, 176, 258n 176
  monthly subsidy to the Earps, 146, 171,
  195, 196, 256n 146

Leussler, Harrison, 178, 185–86, 187,
  209–10, 212, 213, 214, 260n 213
Levy, Harriet Lane, 248n 20
Lewellen, John and Annie, 48, 251n 63
Lewellen, Sarah, 182, 195, 214
Lewis, Alfred Henry, 175, 183
Lewis, Jefferys, 72
Libby, David, 107–8
*Life and Legend of Wyatt Earp, The* (TV
  series), 222
Lindeberg, Jafet, 104, 108
Lockwood, Frank, 155, 200
London, Jack, 94, 113, 157, 223
Los Angeles, California, 9, 136, 138
  Hollenbeck Hotel, 79
  horseracing and, 79
  Josephine Earp's final years in, 205,
  214, 217–18
  Wyatt Earp's arrest in, 142–44
  Wyatt Earp's final years in, 170–73
*Los Angeles Express*, 135
*Los Angeles Herald*, 148
*Los Angeles Times*, 171
  article about Earp and retraction,
  149–50
"Lurid Trails Are Left by Olden-Time
  Bandits" (Scanland), 149

Macartney, Felton, 222
Maddox, Mrs. Lon, 217, 219
Madison Square Garden, New York City,
  138, 174
"manifest destiny," 34
Mansfield, Sadie, 31, 48, 70, 185, 258n
  186
Marcus, Hyman (father), 14, 15, 27, 248n
  14
  death of, 85
Marcus, Moses (grandfather), 15
Marcus, Nathan (brother), 15, 69, 125,
  127, 128, 133, 142

Marcus, Sophia Lewis (mother), 14, 15, 27, 85, 102, 142, 248n 14
  death of, 145
  obituary, 17, 145
Markham, Pauline, 23, 25, 26, 47, 145, 186, 210, 249n 23
  "Pinafore on Wheels" Troupe, 23–26, 145, 186, 210
Marquis, O. H., 227
Martin, Al, 136, 137, 138
Masterson, Bat, 7, 156, 175, 186, 257n 170
  appearance, 72
  description of Wyatt Earp, 33, 149, 211
  dislike of Doc Holliday, 37, 71
  friendship with Wyatt Earp, 71, 73, 78–79, 148–49, 154
  Josephine Earp and, 35, 72
  meets Wyatt Earp, 35
  Stuart Lake and, 170
Masterson, Emma, 72, 73, 183
Mayo, Al, 96, 97, 104
Mayo, Maggie, 97, 102, 104
McCarthy, Lottie and Nellie, 23–24
McCarthy Dancing Academy, 23–24, 210
McKenzie, Alexander, 130, 131–32
McKinley, William, 132
McLaury, Frank, 54, 74
  Gunfight at the O.K. Corral, 55, 56
McLaury, Tom, 53–54
  Gunfight at the O.K. Corral, 55, 56
McLaury, Will, 58, 60, 74
Menckens, Adah Isaac, 20
Miller, Bill, 154–61, 229–30
Miller, Estelle Josephine, 229–30
Mix, Tom, 7, 153, 173–74, 175, 176, 257n 153
Mizner, Wilson, 103, 113, 125, 173, 175
Morey, Jeff, 11
Mudd, Seeley Wintersmith, 256n 143
Muir, John, 93
*My Darling Clementine* (film), 219

Nevitt, Wilfrid and Bessie, 214, 217, 218, 219
New York City
  birth of Josephine Marcus, 3, 5, 15, 247n 5
  conditions of poor in, 15–16
  education of immigrants, 16
  Five Points neighborhood, 15, 17
  Jewish community in, 15, 248n 14, 248n 16
  Jewish newspapers in, 16
  Marcus family in, 15–16, 17, 247n 5
*New York Herald*, 149
*New York Herald Tribune*, 192
*New York Times*, 133, 198
*New York Tribune*, 128
Nome, Alaska, 9, 107–34
  as boomtown, 112, 132–33
  Christmas Day turkey dinners, 118
  claim jumping and litigation, 109
  Clum in, 128, 255n 128
  conditions in, 112, 120–23
  crime in, 123–24
  Earps arrive, 107, 111
  Earps depart, 133–34
  Earps' income in, 132
  Earps' Tombstone friends arrive in, 129–30
  exodus from, 114–15, 133, 254n 114
  gambling and saloon trade in, 106, 118, 124
  gold on the beach, 107, 110–11
  gold rush, 103, 104, 105, 108–9, 121–22
  Jewish community of, 125–26
  Josephine Earp aids storm victims, 131
  Josephine Earp barred from high society, 125
  legal fraud in, 130, 131, 132
  Lucky Baldwin in, 129–30
  mail delivery, 128–29, 255n 128
  Marcus family in, 120, 125, 133

naming of, 108
newspapers, 124–25
Northern Saloon, 106, 113
ordinance against women in saloons, 113–14
population of, 122
prizefighting, 124
red-light district and prostitution, 126–27
Rickard as first mayor, 118
Rickard in, 106, 118
social stratification in, 114, 124–25
storm of 1900, 131, 133
synagogue in, 126
Three Lucky Swedes, 108, 109, 113
winter in, 114–15, 117–18
Wyatt and Josephine Earp, marital tensions in, 126–27
Wyatt and Josephine Earp in, 112–15, 119–34
Wyatt Earp bypassed for deputy sheriff job, 125
Wyatt Earp's brushes with the law in, 127, 128
Wyatt Earp's Dexter Saloon, 113, 117, 122, 124, 126, 130, 132, 133, 255n 133
*Nome Chronicle*, 126
*Nome Daily News*, 127, 128, 131
*Nome Gold Digger*, 126, 130
*Nome Nugget*, 118, 132
Noyes, Arthur, 130, 131–32

Oakland, California, 259n 196
Gertrude Stein and, 21, 248n 21
Josephine Earp in, 137, 146, 179, 190, 195
Lehnhardt mansion in, 74, 85, 137, 142, 145, 195, 217, 252n 74
Lehnhardt's Candies, 141, 145, 146, 195, 215
Wyatt and Josephine Earp in, 142

O'Brian, Hugh, 222
Oddie, Tasker, 137, 148
O.K. Corral
Breakenridge's book about, 169
charges and trial following Gunfight (Spicer hearing), 57–59
as cultural symbol, 3
dead and wounded, 56, 169
Gunfight at, 2–3, 10, 12, 55–56, 140, 221
Gunfight at the O.K. Corral, reenactments, 178, 179
Hollywood and, 151–54
Josephine Marcus and, 3, 6–7, 10, 12
later accounts of, 148–50
lead-up to gunfight, 51–54
naming of corral, 10
versions of gunfight, 3
Ouray, Colorado, 72

Parker, California, 195
Parsons, George, 39, 40, 45, 46, 60–61, 132, 148, 169–70, 189, 193, 200, 202, 250n 40
on cowboys, 43
departure from Tombstone, 67
in Nome, Alaska, 129
on Stilwell killing, 65
Wyatt Earp and, 67–68
Peabody, Endicott, 7, 39, 40, 65, 166
on "Cowboy and Earp Feud," 60
on crime in Tombstone, 43
departure from Tombstone, 67
on Wyatt Earp, 67
Peoria, Illinois, 36, 211
Phoenix, Arizona, 28
*Pioneer Days in Arizona* (Lockwood), 200
Pitkin, Frederick, 71, 252n 71
Planet Earp, 5
Posen, Prussia, 14, 248n 14, 248n 20
Prescott, Arizona, 25–26, 48
Virgil Earp as deputy U.S. marshal, 38

Prohibition, 144, 158, 175
prostitution, 7, 36, 201
  Bessie Earp and, 36, 42
  Josephine Earp suspected of, 48–49,
    186, 209, 258n 186, 260n 209
  Mattie Earp and, 84–85
  in Nome, Alaska, 126–27
  Sally Haspel and, 36, 211
  in San Diego, 82
  in Tombstone, 47–49, 250n 49, 258n
    186, 260n 209
  Wyatt Earp and brothels, 36, 82, 127,
    137, 211

railroads
  "big red" train, Los Angeles to San
    Bernadino, 180
  San Francisco–Los Angeles line com-
    pleted (1876), 21
  transcontinental railroad (1869), 18
Raine, William MacLeod, 167, 168, 169,
  192
Rampart, Alaska, 96–104, 116, 125
  Earps' home in, 97–98, 254n 97
  Josephine Earp's happiness in,
    100–103
  prospecting and gold in, 97–98
  town dances, 101
Randall, George M., 133
Rickabaugh, Lou, 86
Rickard, Tex, 101, 103–4, 106, 111, 113,
  118, 124, 134, 138, 174
  as fight promoter, 124, 138
Ringo, Johnny, 168
Roosevelt, Franklin Delano, 40
Rosenthal, Toby, 20
"Rose Still Grows beyond the Wall,
  The" (Frank), 218

*Saga of Billy the Kid, The* (Burns), 161,
  163

*Saidie* (ship), 111
St. John, Adela Rogers, 174
*Saint Johnson* (Burnett), 184, 187, 188
Salt Lake City, Utah, 71–72
San Antonio, Texas, 208
San Bernardino, California, 35
  Earp's common-law wife Mattie in, 63,
    68
  Earp's parents in, 38, 63, 249n 38
San Diego, California, 9, 125
  as boomtown, 76
  Coronado Island, 78, 147–48
  Earp brothers in, 77
  "Golden Poppy" brothel, Wyatt Earp's
    ties to, 82
  horseracing, 79–82
  moral reforms in, 83
  Oyster Bar and Gambling Hall, 77, 82
  real estate market collapse, 83
  Wyatt and Bat Masterson law enforce-
    ment assignment, 78–79
  Wyatt and brothels in, 82
  Wyatt and Josephine Earp in, 76–84,
    252n 82
  Wyatt Earp's business ventures in,
    77–83
*San Diego Union*, 64, 83
  Clara Brown writing for, 39–40
San Francisco, California, 125
  Academy of Music, 20
  affluence in, 16, 17, 22, 74
  art and culture in, 20, 21
  Baldwin Hotel, 79, 91, 130
  commerce and economic conditions in,
    19th century, 16–17
  depression of 1876, 21–22
  earthquake, 1868, 18
  earthquake and fire, 1906, 141–42, 173,
    182
  Eureka Benevolent Association, 19
  Goodfellow's Grotto Café, 90

Hills of Eternity Cemetery, Colma, California,1, 4, 85, 145–46, 176, 218, 221, 258n 176

Israel Benjamin's description of, 18–19

Jewish community of, 4, 16–17, 19, 20–21, 142, 248n 16

Jewish newspapers in, 16

Jewish women, careers of, 20–21

Lucky Baldwin's hotel, 79

Marcus family in, 14, 18–21, 69–71, 74, 82, 102

social stratification in, 19, 248n 19

theaters and acting companies, 20, 23

Wyatt and Josephine Earp in, 117

Wyatt and Josephine Earp in, 1890–1897, 85–88, 90–91

Wyatt and Josephine Earp's home in, 86

Wyatt Earp and fight-fixing scandal, 90–91, 143, 253n 91

Wyatt Earp buried in, 1, 4, 176, 218, 221, 258n 176

Wyatt Earp's accident in, 93

Wyatt Earp's horseracing and stables, 85–86

*San Francisco Chronicle*, 91, 119, 192

*San Francisco Examiner*, 90

*Saturday Evening Post*, 161, 162, 164, 191

Scanland, J. M., 149

Schieffelin, Ed, 29, 32, 96

Seattle, Washington, 116–17, 125, 254n 119

Alaska gold rush and, 100, 116–17, 254n 100

exodus to Nome from, 119

as "Nome crazy," 116

Wyatt and Josephine Earp in, 117, 254n 119

*Seattle Post-Intelligencer*, 124, 128

Sharkey, Tom, 90–91, 124, 253n 91

Sherman, William Tecumseh, 66

Short, Luke, 73

Shurtleff, Dr. Fred, 174, 175

Sieber, Al, 25

"Significance of the Frontier in American History" (Turner), 87, 253n 87

Sinclair, Addie, 216, 260n 216

Skagway, Alaska, 93

*Skookum* (ship), 131

Smith, Jefferson Randolph "Soapy," 93

Spence, Mariette "Maria" Duarte, 28, 30, 42, 164, 251n 63

testifies against husband, 62–63

Spence, Pete, 42

as Morgan killer suspect, 62, 63, 251n 63

as stagecoach robbery suspect, 52

Spicer, Judge Wells, 59, 60

*Spoilers, The* (Beach), 107, 125, 134

Spolidoro, Grace Welsh, 4, 140–41, 176

*Squaw Man, The* (film), 151

*Brixon* (ship), 95

*Roanoke* (ship), 134

Stegner, Wallace, 8

Stein, Gertrude, 21, 248n 20, 248n 21

Stilwell, Frank

as stagecoach robbery suspect, 52

Wyatt Earp kills, 64, 154

St. Michael, Alaska, 103, 104

Wyatt and Josephine Earp in, 104–7

Strauss, Levi, 16

*Sunset Trail, The* (Lewis), 183, 175

*Suppressed Murder of Wyatt Earp* (Boyer), 230

Tabor, Baby Doe, 72, 252n 72

Tefertiller, Casey, 228–29, 234

Thompson, Lydia, 23

Tijuana, Mexico, 77

Toklas, Alice B., 248n 20

*Tombstone* (film), 234

Tombstone, Arizona, 9, 13, 182. *See also*
  Earp, Josephine "Sadie" Marcus;
  Earp, Wyatt
beef demand in, 44–45
Behan in, 31
Bird Cage Theatre, 26, 178, 209, 224
books about, 167–69, 191–94, 199–200,
  228
Citizens Safety Committee, 53, 56
city in 21st century, 9–10
city in 1880, 28–30
city in 1881, 42–49
Clantons in, 45
conventional citizens of, 39
Cosmopolitan Hotel, 30, 45, 60, 61
cowboys' funeral after Gunfight, 56–57
cowboys in, 36, 43, 44–45, 169, 179
crime in, 43, 44–45, 50–51, 53
demimonde of (and lesbian reference),
  39, 40, 47–48, 250n 40
demise of, 1882-1886, 66–68
doctor in, 48
"downtown" of, 41
Earp brothers in, 10, 209
Earp-cowboy war, 59–64
Earp family arrives, 24, 32, 37, 38
Earp family departure, 66
Earp's Vendetta Ride, 64–66, 148, 168
eyewitness accounts of, 39–41
factions in, 10, 50–51
fire department formed in, 45–46
fire of 1881, 45
first mayor, John Clum, 29
"Forty-Dollar Sadie," 48, 185–86
General Sherman visits, 66
Grand Hotel, 30, 32, 45, 60, 251n 60
Gunfight at the O.K. Corral, 2–3, 10,
  12, 55–56, 140, 221
Helldorado festival, 9–10, 177–79
Hollywood and, 152, 154
influx of people to, 29–30

Jewish community of, 42
Josephine Earp returns, 1937, 208–9
Josephine Marcus arrives, 1880, 7,
  13–14, 247n 13
Josephine Marcus departs, 7, 63, 251n
  63
Josephine Marcus in acting troupe,
  1879, 24–25
killing of Morgan Earp, 61–62, 251n 61,
  251n 63
lawmen as tax collectors, 39
Lotta Crabtree estate case, 166
Markham "Pinafore on Wheels"
  Troupe in, 24–25
McLaurys in, 45
mining in, 29, 42–43
newspapers, 51
Oriental Saloon, 59, 60, 171
origins of, 29
politics in, 32, 44, 50–51
prostitution in, 47–49, 144, 209, 250n
  49, 258n 186, 260n 209
ratio of men to women, 49
Schieffelin Hall, 61, 178, 251n 61
shooting of Virgil Earp, 61
social stratification in, 10, 39, 40
support for Earps following Gunfight,
  57
television series about Earp and, 222
Union News Depot, 42, 49
weather of, 38, 50
Wells Fargo Company office, 29
women in, 41, 45–46, 49
Wyatt Earp's business ventures in,
  38–39
*Tombstone: An Iliad of the Southwest*
  (Burns), 33, 161–62, 163, 168
*Tombstone Epitaph*, 51, 63, 117, 128, 144,
  177, 208, 228
Tombstone *Nugget*, 51, 169
*Tombstone Vendetta* (Boyer), 233

Tonopah, Nevada, 135–38
  Northern Saloon, 137, 138
  Wyatt Earp's business ventures in, 137
*Tonopah Bonanza*, 137
Tonopah Mining Company, 137
transportation/travel
  to Alaska's gold fields, 93
  railroads, 13, 21
  San Francisco-Los Angeles railroad
    completed (1876), 21
  boat travel between Nome to Seattle,
    115, 119–20
  stagecoach, 13, 28
  transcontinental railroad (1869), 18
  wagontrains, 34–35
  westward via Isthmus of Panama,
    17–18
  Tucson, Arizona, 28
  killing of Frank Stilwell, 63
  Old Pueblo Club, 168
Turner, Frederick Jackson, 87, 253n 87
Twain, Mark, 87, 223

Unalaska, Alaska, 95, 120
*Under Cover for Wells Fargo* (Dodge), 200
Unga, Alaska, 132

Van Buren, Martin, 10
Vawter, Cornelius, 106, 125, 130, 131,
  132
Vawter, Sarah, 106, 125, 131, 132
Vidal, California, 140, 165, 173, 180, 194
*Virginian, The* (Wister), 88

Walsh, Raoul, 152, 153
Waters, Frank, 206, 207, 209, 228–29
Waters, Naomi, 207

*Weekly Arizona Miner*, 25
Wiener, Aaron, 69, 71, 85, 142, 248n 19
Wiener, Rebecca Lewis (half-sister), 19,
  20, 69, 85, 142, 147, 248n 19
Wells, Alice Earp, 200, 225
Wells Fargo Bank, 16
Wells Fargo Company
  attack on Benson stagecoach, 51–52
  Morgan Earp and, 28
  reward for Benson killers, 52
  Tombstone office, 29
  Virgil Earp and, 71
  Wyatt Earp and, 38–39, 52, 70, 75, 149
Welsh, Charlie, 140, 158, 175, 255n 120
Welsh, Christenne, 180
Welsh, Grace, 180
Wichita, Kansas, 36
Wilcox, Arizona, 127
Wister, Owen, 87–88
women,
  biographies of, 6
  life of, in Tombstone, 41, 45–46
  in Nome, Alaska, 113–14
  as prostitutes, 7, 47–49
  as "shrill or strident," 6
  in wagontrains, 34–35, 249n 34
  Western frontier life and, 7
  in Western history, 6, 164
  working, on the frontier, 47
Wrangell, Alaska, 93
Wurtzel, Sol, 215
*Wyatt Earp* (film), 234
*Wyatt Earp* (ship), 197–99
*Wyatt Earp: The Life Behind the Legend*
  (Tefertiller), 234

Yuma, Arizona, 89, 92

# | About the Author

Ann Kirschner is University Dean of Macaulay Honors College at the City University of New York. She began her career as a lecturer in Victorian literature at Princeton University, where she earned her Ph.D. A writer of wide-ranging interests, she is the author of *Sala's Gift* and an innovator in digital media and education. She lives in New York City with her family.